R Programming for
Bioinformatics

Chapman & Hall/CRC
Computer Science and Data Analysis Series

The interface between the computer and statistical sciences is increasing, as each discipline seeks to harness the power and resources of the other. This series aims to foster the integration between the computer sciences and statistical, numerical, and probabilistic methods by publishing a broad range of reference works, textbooks, and handbooks.

SERIES EDITORS
David Blei, Princeton University
David Madigan, Rutgers University
Marina Meila, University of Washington
Fionn Murtagh, Royal Holloway, University of London

Proposals for the series should be sent directly to one of the series editors above, or submitted to:

Chapman & Hall/CRC
4th Floor, Albert House
1-4 Singer Street
London EC2A 4BQ
UK

Published Titles

Bayesian Artificial Intelligence
Kevin B. Korb and Ann E. Nicholson

Computational Statistics Handbook with MATLAB®, Second Edition
Wendy L. Martinez and Angel R. Martinez

Pattern Recognition Algorithms for Data Mining
Sankar K. Pal and Pabitra Mitra

Exploratory Data Analysis with MATLAB®
Wendy L. Martinez and Angel R. Martinez

Clustering for Data Mining: A Data Recovery Approach
Boris Mirkin

Correspondence Analysis and Data Coding with Java and R
Fionn Murtagh

Design and Modeling for Computer Experiments
Kai-Tai Fang, Runze Li, and Agus Sudjianto

Introduction to Machine Learning and Bioinformatics
Sushmita Mitra, Sujay Datta, Theodore Perkins, and George Michailidis

R Graphics
Paul Murrell

R Programming for Bioinformatics
Robert Gentleman

Semisupervised Learning for Computational Linguistics
Steven Abney

Statistical Computing with R
Maria L. Rizzo

Computer Science and Data Analysis Series

R Programming for Bioinformatics

Robert Gentleman

Fred Hutchinson Cancer Research Center

Seattle, Washington, U.S.A.

CRC Press
Taylor & Francis Group
Boca Raton   London   New York

CRC Press is an imprint of the
Taylor & Francis Group, an **informa** business
A CHAPMAN & HALL BOOK

Chapman & Hall/CRC
Taylor & Francis Group
6000 Broken Sound Parkway NW, Suite 300
Boca Raton, FL 33487-2742

© 2009 by Taylor & Francis Group, LLC
Chapman & Hall/CRC is an imprint of Taylor & Francis Group, an Informa business

---

Library of Congress Cataloging-in-Publication Data

Gentleman, Robert, 1959-
    R programming for bioinformatics / Robert Gentleman.
        p. cm. -- (Chapman & Hall/CRC computer science and data analysis series)
        Bibliographical references (p.) and index.
    ISBN 978-1-4200-6367-7
    1. Bioinformatics. 2. R (Computer program language) I. Title. II. Series.

QH324.2.G46 2008
572.80285'5133--dc22   2008011352

Visit the Taylor & Francis Web site at
http://www.taylorandfrancis.com

and the CRC Press Web site at
http://www.crcpress.com

To Tanja, Sophie and Katja

Contents

xii

Chapter 1

Introducing R

1.1 Introduction

The purpose of this monograph is to provide a reference for scientists and programmers working on problems in bioinformatics and computational biology. It may also appeal to programmers who want to improve their programming skills or programmers who have been working in bioinformatics and computational biology but are familiar with languages other than R. A reasonable level of programming skill is presumed as is some familiarity with some of the basic tasks that need to be carried out in bioinformatics. We concentrate on programming tools and there is no discussion of either graphics or of the multitude of software for fitting models or carrying out machine learning. Reasonable coverage of these topics would result in a much longer monograph and to some extent they are orthogonal to our purpose.

Bioinformatics blossomed as a scientific discipline in the 1990s when a number of technological innovations appeared that revolutionized biology. Suddenly, data on the complete genomic sequence of many different organisms were available, microarrays could measure the abundance of tens of thousands of mRNA species, and other arrays and technologies made it possible to study protein interactions and many other cellular processes at the molecular level. Basically, biology moved from a small data discipline to one with large complex data sets, virtually overnight.

Faced with these sudden challenges, scientific programmers grabbed whatever tools were available and made use of them to help address some of the many problems. Perl was perhaps the most widely used and it remains a dominant player to this date. Other popular programming languages such as Java and Python are also used.

R is an implementation of the S language (Becker et al., 1988; Chambers and Hastie, 1992; Chambers, 1998). S has been a favorite tool for statisticians and data analysts since the early 1980s when John Chambers and colleagues started to release versions of it from Bell Labs. It is now becoming one of the most widely used software tools for bioinformatics. This is mainly due to its flexibility and data handling and modeling capabilities. Some of these have been exposed through the Bioconductor Project (Gentleman et al., 2004) but many users simply find it a useful tool for doing analyses. However, our experience is that it is easy to write inefficient programs, and often the basic programming idioms are missed or ignored.

In Chapter 2 we discuss the general properties of the R language and some of the unique aspects of programming in it. In Chapter 3 we discuss object-oriented programming in R. The paradigm is quite different and may take some getting used to, but like all object-oriented systems, mastering these topics is essential to writing good maintainable software. Then Chapter 4 discusses methods for getting data in and out, for interacting with databases and includes a discussion of XML, SOAP and other data mark-up and web-services languages and tools. Chapter 5 discusses different aspects of string handling and manipulations, including many of the standard sequence similarity tools that play a prominent role in computational biology. In Chapter 6 we consider interacting with foreign languages, primarily on C, but we also consider FORTRAN, Perl and Python. In Chapter 7 we describe how to write your own software packages that can be used locally or distributed more broadly. Finally we finish with Chapter 9, which discusses debugging and profiling of R code.

R comes with a substantial amount of documentation. Specifically there are five manuals: An Introduction to R, The R Language Definition, R Installation and Administration, Writing R Extensions, and R Data Import and Export. We will draw on material in these manuals throughout this monograph, and readers who want more detail or alternative examples should consult them. We will rely most on the Writing R Extensions Manual, which we abbreviate to R Extensions. R News is a good source of information on R packages and on aspects of the language written at an accessible level. Readers are encouraged to browse the back issues for more information on topics that are just touched on in this volume. Venables and Ripley (2000) is another reference for programming in the S language, as is Chambers (2008).

1.2   Motivation

There are many good reasons to prefer R to other languages for scientific computation. The existence of a substantial collection of good statistical algorithms, access to high-quality numerical routines, and integrated data visualization tools are perhaps the most obvious ones. But as we have been trying to show through the Bioconductor Project (www.bioconductor.org), there are many more.

Reproducibility is an essential part of any scientific investigation, but to date very little attention has been paid to this topic. Our efforts are R-based (Gentleman, 2005) and make use of the Sweave system (Leisch, 2002). Indeed, as we discuss later, this entire book has been written so that every example is reproducible on the reader's machine. The ability to integrate text and software into a single document greatly facilitates the writing of scientific papers and helps to ensure that all figures, tables and facts are based on the same data, and are essentially reproducible by the reader.

A second strong motivation for using R is its ability to interoperate with many other languages. Algorithms that have been written in another language seldom need to be reimplemented for use in R. Typically one need merely write a small amount of interface code and the routines can be accessed from within R (this is described in Chapter 6). This approach also helps to ensure maximal code reuse.

And finally, R supports the creation and use of self-describing data structures. In the Bioconductor Project we have relied heavily on this capability in our design and use of the *ExpressionSet* class. This data structure is designed to hold the output of a microarray experiment as well as detailed information on the experimental design, other covariates that are available for the samples that were run and links to information on the genes that correspond to the spots on the array. While this has been successful in that context, we have reused this data structure with similar benefits for other data types such as those that arise in proteomic studies (the **PROcess** package) and flow cytometry (the **flowCore** package).

1.3 A note on the text

This monograph was written using the Sweave system (Leisch, 2002), which is a tool that allows authors to integrate text (using LaTeX) and computer code for the R language. Hence, all examples are reproducible by the reader and readers can obtain complete source code (but not the text) on a per-chapter basis from the web site for this monograph. There are a number of exercises given and solutions for some of them are available in the online supplements.

The examples themselves are often shown integrated into the text of the chapters. Not all code is displayed; in many cases preliminary computations, loading of libraries and other mundane tasks are not displayed in the text version; they are included in the R code for the chapters. Any example that relies on simulation or the use of a random number generator will have a call to the `set.seed` function as a preliminary command. The sole reason for this is to ensure reproducibility of the output on the user's machine.

In cases where the code is intended to signal an error, the call is enclosed in either a call to `try` or more often in a call to `tryCatch`. This is done because any error signaled by R interrupts the Sweave process and causes typesetting to fail. Details on the behavior of `try` and `tryCatch` can be found in Section 2.11.

Markup is used to distinguish some entities. For example, functions are marked up like `mean`, R packages using **Biobase**, function arguments, `myarg`, R classes with *ExpressionSet*, and R objects using `x`. When R prints a value that corresponds to a vector, some indexing information is provided. In the example below, we print a vector of integers from 1 to 30. The first thing printed is [1], which indicates that the first value on that line is the first value in the vector, and on the second printed line the [18] indicates that the first value in that line corresponds to the 18th element of the vector. The grey background is used for all code examples that were processed by the Sweave system.

```
> 1:30

 [1]  1  2  3  4  5  6  7  8  9 10 11 12 13 14 15 16 17
[18] 18 19 20 21 22 23 24 25 26 27 28 29 30
```

It is essential that the reader follow along and experiment with some of the examples given, so two basic strategies are advised. First, make use of the help system, either by constructs such as `help("[")` or the shorthand, equivalent, `?"["`. Many special functions and symbols need to be quoted. All help pages should have examples and these can be run using the function `example`, e.g., `example("[")`). The second basic strategy is to investigate the code itself, and for this purpose `get` is most useful; for example, try `get("mode")` and see if you can better understand how it works.

1.4 Acknowledgments

Many people have contributed, both directly and indirectly, to the creation of this book. Both the R and Bioconductor development teams have contributed substantially to my understanding, and many members of those projects have provided examples, clarified misunderstandings and provided a rich environment in which to discuss relevant issues. Members of my research group have contributed to many aspects; in particular, J. Gentry, S. DebRoy, H. Pagés, M. Morgan, N. Li, T.-Y. Liu, M. Carlson, P. Aboyoun, D. Sarkar, F. Hahne and S. Falcon have contributed many ideas, examples and helped clarify issues. Wolfgang Huber and Vincent Carey made extensive comments and recommendations. All errors remain my own and I will attempt to remedy those that are found and reported, in a timely fashion.

Chapter 2

R Language Fundamentals

2.1 Introduction

In this chapter we introduce the basic language data types and discuss their capabilities and structures. Then topics such as flow-control, iteration, subsetting and exception handling will be presented. R directly supports two different object-oriented programming (OOP) paradigms, which are discussed in detail in Chapter 3. Many operations in R are vectorized, and understanding and using vectorization is an essential component of becoming a proficient programmer.

The R language was primarily designed as a language for data manipulation, modeling and visualization, and many of the data structures reflect this view. However, R is itself a full-fledged programming language, with its own idioms – much like any other programming language. In some ways R can be considered as a functional programming language, although it is not purely functional. R supports a form of lexical scope that provides a useful paradigm for encapsulating computations.

R is an implementation of the S language (Becker et al., 1988; Chambers and Hastie, 1992; Chambers, 1998). There is another commercial implementation available from Insightful Corporation, called S-PLUS. The two implementations are quite similar, and much of the material covered here can be used in either. However, there are many R-specific extensions that are used in this monograph and users of R are our intended audience.

2.1.1 A brief introduction to R

We presume a reasonable familiarity with R but there are a few points that will help to clarify some of the discussion. When R is started, a workspace is created and that workspace is where the user creates and manipulates variables. This workspace is an environment, and an environment is a set of bindings of names, or symbols, to values. The top-level workspace can be accessed through its name, which is `.GlobalEnv`.

Assignment of value to a variable is generally done with either the = (equals) character, or a special symbol that is the concatenation of less than and minus, `<-`. Assignment creates a binding between a symbol and a value, in a particular environment. Removal of bindings is done with the function `rm`. In the next code chunk, we create a symbol x and assign to it the value 10. We then create a second symbol and assign the same value as x has.

```
> x = 10
> y = x
```

The value associated with y is a copy of the value associated with x, and changes to x do not affect y.

The semantics of `rm(x)` are that the association between x and its value is broken and the symbol x is removed from the environment, but nothing is done to the value that x referred to. If this value can be accessed in other ways, it will remain available. We provide an example in Section 2.2.4.3.

Valid variable names, sometimes referred to as *syntactic names*, are any sequence of letters, digits, the period and the underscore, but they cannot begin with a digit or the underscore. If they begin with a period, the second character cannot be a digit. Variable names that violate these rules must be quoted (see the Quotes manual page) and the preferred quote is the backtick.

```
> `_foo` = 10
> "10:10" = 20
> ls()

[1] "10:10"    "Rvers"   "_foo"      "basename"
[5] "biocUrls" "repos"   "x"         "y"
```

2.1.2 Attributes

Attributes can be attached to any R object except `NULL` and they are used quite extensively. Attributes are stored, by name, in a list. All attributes can be retrieved using `attributes`, or any particular attribute can be accessed or modified using the `attr` function. Attributes can be used by programmers to attach any sort of information they want to any R object. R uses attributes for many things; the S3 class system is based largely on attributes, dimensions of arrays, and names on vectors, to name but a few.

In the code below, we attach an attribute to x and then show how the printing of x changes to reflect the fact that it has an attribute.

```
> x = 1:10
> attr(x, "foo") = 11
> x
```

```
[1]  1  2  3  4  5  6  7  8  9 10
attr(,"foo")
[1] 11
```

2.1.3 A very brief introduction to OOP in R

In order to fully explain some of the concepts in this chapter, the reader will need a little familiarity with the basic ideas in object oriented-programming (OOP), as they are implemented in R. A more comprehensive treatment of these topics is given in Chapter 3. There are two components: one is a class system that is used to define the class of different objects, and the second is the notion of a generic function with methods. R has two OOP systems: one is referred to as S3, and it mainly supports generic functions; the other is referred to as S4, and it has support for classes as well as generic functions, although these are somewhat different from the S3 variants. We will only discuss S3 here.

In S3, the class system is very lax, and one creates an object (typically called an instance) from a class by attaching a `class` attribute to any R object. As a result, no checking is done, or can easily be done, to ensure common structure of different instances of the same class. A generic function is essentially a dispatching mechanism, and in S3 the dispatch is handled by concatenating the name of the generic function with that of the class. An example of a generic function is `mean`.

```
> mean

function (x, ...)
UseMethod("mean")
<environment: namespace:base>
```

The general form of a generic function, as seen in the example above, is for a single expression, which is the call to `UseMethod`, which is the mechanism that helps to dispatch to the appropriate method. We can see all the defined methods for this function using the `methods` command.

```
> methods("mean")

[1] mean.Date       mean.POSIXct    mean.POSIXlt
[4] mean.data.frame mean.default    mean.difftime
```

And see that they all begin with the name **mean**, then a period. When the function **mean** is called, R looks at the first argument and determines whether or not that argument has a class attribute. If it does, then R looks for a function whose name starts with **mean.** and then has the name of the class. If one exists, then that method is used; and if one does not exist, then **mean.default** is used.

2.1.4 Some special values

There are a number of special variables and values in the language and before embarking on data structures we will introduce these. The value **NULL** is the null object. It has length zero and disappears when concatentated with any other object. It is the default value for the elements of a list.

```
> length(NULL)

[1] 0

> c(1, NULL)

[1] 1

> list("a", NULL)

[[1]]
[1] "a"

[[2]]
NULL
```

Since R has its roots in data analysis, the appropriate handling of missing data items is important. There are special missing data values for all atomic types and these are commonly referred to by the symbol **NA**. And similarly there are special functions for identifying these values, such as **is.na**, and many modeling routines have special methods for dealing with missing values. It is worth emphasizing that there is a distinct missing value (NA) for each basic type and these can be accessed through constants such as **NA_integer_**.

```
> typeof(NA)

[1] "logical"

> as.character(NA)
```

```
[1] NA

> as.integer(NA)

[1] NA

> typeof(as.integer(NA))

[1] "integer"
```

Note that the character string formed by concatenating the characters N and A is not a missing value.

```
> is.na("NA")

[1] FALSE
```

The appropriate representation of values such as infinity and *not a number* (NaN) is provided. There are accompanying functions, `is.finite`, `is.infinite` and `is.nan`, that can be used to determine whether a particular value is one of these special values. All mathematics functions should deal with these values appropriately, according to the ANSI/IEEE 754 floating-point standard.

```
> y = 1/0
> y

[1] Inf

> -y

[1] -Inf

> y - y

[1] NaN
```

2.1.5 Types of objects

An important data structure in R is the vector. Vectors are ordered collections of objects, where all elements are of the same type. Vectors can be of any length (including zero), up to some maximum allowable, which is determined by the storage capabilities of the machine being used. Vectors typically

represent a form of contiguous storage (character vectors are an exception).
R has six basic vector types: logical, integer, real, complex, string (or character) and raw. The type of a vector can be queried by using one of the three
functions `mode`, `storage.mode` or `typeof`.

```
> typeof(y)

[1] "double"

> typeof(is.na)

[1] "builtin"

> typeof(mean)

[1] "closure"

> mode(NA)

[1] "logical"

> storage.mode(letters)

[1] "character"
```

There are also a number of predicate functions that can be used to test
whether a value corresponds to one of the basic vector types. The code chunk
below demonstrates the use of several of the predicate functions available.

```
> is.integer(y)

[1] FALSE

> is.character(y)

[1] FALSE

> is.double(y)

[1] TRUE

> is.numeric(y)

[1] TRUE
```

Exercise 2.1
What does the `typeof` `is.na` mean? Why is it different from that of `mean`?

2.1.6 Sequence generating and vector subsetting

It is also helpful to discuss a couple of functions operators before beginning the general discussion, as they will help make the exposition easier to follow. Some of these, such as the subsetting operator `[`, we will return to later for a more complete treatment.

The colon, `:`, indicates a sequence of values, from the number that is to its left, to the number on the right, in steps of 1. We will also need to make some use of the subset operator, `[`. This operator takes a subset of the vector it is applied to according to the arguments inside the square brackets.

```
> 1:3

[1] 1 2 3

> 1.3:3.2

[1] 1.3 2.3

> 6:3

[1] 6 5 4 3

> x = 11:20
> x[4:5]

[1] 14 15
```

These are just ordinary functions, and one can invoke them as if they are. The usual infix notation, with the `:` between the lower and upper bounds on the sequence, may lead one to believe that this is not an ordinary function. But that is not true, and one can also invoke this function using a somewhat more standard notation, `":"(2,4)`. Quotes are needed around the colon to ensure it is not interpreted in an infix context by the parser.

Exercise 2.2
Find help for the colon operator; what does it do? What is the type of its return value? Use the predicate testing functions to determine the storage mode of the expressions `1:3` and `1.3:4.2`.

2.1.7 Types of functions

This section is slightly more detailed and can be skipped. In R there are basically three types of functions: builtins, specials and closures. Users can only create closures (unless they want to modify the internals of R), and these are the easiest functions to understand since they are written in R. The other two types of functions are interfaces that attempt to pass the calculations down to internal (typically C) routines for efficiency reasons. The main difference between the two types of internal functions is whether or not they evaluate their arguments; specials do not. More details on the internals of R are available in the R Language Definition (R Development Core Team, 2007b).

2.2 Data structures

2.2.1 Atomic vectors

Atomic vectors are the most basic of all data structures. An atomic vector contains some number of values of the same type; that number could be zero. Atomic vectors can contain integers, doubles, logicals or character strings. Both complex numbers and raw (pure bytes) have atomic representations (see the R documentation for more details on these two types). Character vectors in the S language are vectors of character strings, not vectors of characters. For example, the string `"super"` would be represented as a character vector of length one, not of length five (for more details on character handling in R, see Chapter 5). A `dim` attribute can be added to an atomic vector to create a matrix or an array.

```
> x = c(1, 2, 3, 4)
> x

[1] 1 2 3 4

> dim(x) = c(2, 2)
> x

     [,1] [,2]
[1,]    1    3
[2,]    2    4

> typeof(x)

[1] "double"
```

```
> y = letters[1:10]
> y

 [1] "a" "b" "c" "d" "e" "f" "g" "h" "i" "j"

> dim(y) = c(2, 5)
> y

     [,1] [,2] [,3] [,4] [,5]
[1,] "a"  "c"  "e"  "g"  "i"
[2,] "b"  "d"  "f"  "h"  "j"

> typeof(y)

[1] "character"
```

A logical value is either TRUE, FALSE or NA.

The elements of a vector can have names, and a matrix or array can have names for each of its dimensions. If a dim attribute is added to a named vector, the names are discarded but other attributes are retained (and dim is added as an attribute).

Vectors can be created using the function c, which is short for concatenate. Vectors for a particular class can be created using the functions numeric, double, character integer or logical; all of these functions take a single argument, which is interpreted as the length of the desired vector. The returned vector has initial values, appropriate for the type.

The function seq can be used to generate patterned sequences of values. There are two variants of seq that can be very efficient: seq_len that generates a sequence from 1 to the value provided as its argument, and seq_along that returns a sequence of integers of the same length as its argument. If that argument is of zero length, then a zero length integer vector is returned, otherwise the sequence starts at 1.

The different random number generating functions (e.g., rnorm, runif) can be used to generate random vectors. sample can be used to generate a vector sampled from its input. Notice in the following example that the result of typeof(c(1, 3:5)) is "double", whereas typeof(c(1, "a")) is "character". This is because all elements of a vector must have the same type, and R coerces all elements of c(1, "a") to character.

```
> c(1, 3:5)

[1] 1 3 4 5

> c(1, "c")
```

```
[1] "1" "c"

> numeric(2)

[1] 0 0

> character(2)

[1] "" ""

> seq(1, 10, by = 2)

[1] 1 3 5 7 9

> seq_len(2.2)

[1] 1 2

> seq_along(numeric(0))

integer(0)

> sample(1:100, 5)

[1] 59 89 49 66 10
```

S regards an array as consisting of a vector containing the array's elements, together with a dimension (or `dim`) attribute. A vector can be given dimensions by using the functions `matrix` (two-dimensional data) or `array` (any number of dimensions), or by directly attaching them with the `dim` function. The elements in the underlying vector correspond to the elements of the array. For matrices, the first column is stored first, followed by the second column and so on.

Array extents can be named by using the `dimnames` function or the `dimnames` argument to `matrix` or `array`. Extent names are given as a list, with each list element being a vector of names for the corresponding extent.

Exercise 2.3
Create vectors of each of the different primitive types. Create matrices and arrays by attaching dim *attributes to those vectors. Look up the help for* dimnames *and attach dimnames to a matrix with two rows and five columns.*

2.2.1.1   Zero length vectors

In some cases the behavior of zero length vectors may seem surprising. In Section 2.6 we discuss vectorized computations in R and describe the rules that apply to zero length vectors for those computations. Here we describe their behavior in other settings.

Functions such as `sum` and `prod` take as input one or more vectors and produce a value of length one. It is helpful if simple rules, such as `sum(c(x,y)) = sum(x) + sum(y)`, hold. Similarly for `prod` we expect `prod(c(x,y)) = prod(x)*prod(y)`. For these to hold, we require that the sum of a zero length vector be zero and that the product of a zero length vector be one.

```
> sum(numeric())

[1] 0

> prod(numeric())

[1] 1
```

For other mathematical functions, such as `gamma` or `log`, the same logic suggests that these functions should return a zero length result when invoked with an argument of zero length.

2.2.2 Numerical computing

One of the strengths of R is its various numerical computing capabilities. It is important to remember that computers cannot represent all numbers and that machine computation is not identical to computation with real numbers. Readers unaware of the issues should consult a reference on numerical computing, such as Thisted (1988) or Lange (1999) for more complete details or Goldberg (1991). The issue is also covered in the R FAQ, where the following information is provided.

> The only numbers that can be represented exactly in R's numeric type are (some) integers and fractions whose denominator is a power of 2. Other numbers have to be rounded to (typically) 53 binary digits accuracy. As a result, two floating point numbers will not reliably be equal unless they have been computed by the same algorithm, and not always even then.

And a classical example of the problem is given in the code below.

```
> a = sqrt(2)
> a * a == 2

[1] FALSE
```

```
> a * a - 2

[1] 4.440892e-16
```

The numerical characteristics of the computer that R is running on can be obtained from the variable named .Machine. These values are determined dynamically. The manual page for that variable provides explicit details on the quantities that are presented.

The function all.equal compares two objects using a numeric tolerance of .Machine$double.eps^0.5. If you want much greater accuracy than this, you will need to consider error propagation carefully.

Exercise 2.4
What is the largest integer that can be represented on your computer? What happens if you add one to this number? What is the smallest negative integer that can be represented?

2.2.3 Factors

Factors reflect the S language's roots in statistical application. A factor is useful when a potentially large collection of data contains relatively few, discrete *levels*. Such data are usually referred to as a categorical variable. Examples include variables like sex, e.g., male or female. Some factors have a natural ordering of the levels, e.g., low, medium and high, and these are called ordered factors. While one can often represent factors by integers directly, such practice is not recommended and can lead to hard to detect errors. Factors are generally used, and are treated specially, in different statistical modeling functions such as lm and glm. Factors are not vectors and, in particular, is.vector returns FALSE for a factor.

A factor is represented as an object of class *factor*, which is an integer vector of codes and an attribute with name levels. In the code below, we first set the random seed to ensure that all readers will get the same values if they run the code on their own machines.

```
> set.seed(123)
> x = sample(letters[1:5], 10, replace = TRUE)
> y = factor(x)
> y

 [1] b d c e e a c e c c
Levels: a b c d e

> attributes(y)
```

```
$levels
[1] "a" "b" "c" "d" "e"

$class
[1] "factor"
```

The creation of factors typically either happens automatically when reading data from disk, e.g., `read.table` does automatic conversion, or by converting a character vector to a factor through a call to the function `factor` unless the option `stringsAsFactors` has been set to `FALSE`. When `factor` is invoked, the following algorithm is used. If no `levels` argument is provided, then the levels are assigned to the unique values in the first argument, in the order in which they appear. Values provided in the `exclude` argument are removed from the supplied `levels` argument. Then if `x[i]` equals the jth value in the `levels` argument, the ith element of the result is j. If no match is found for `x[i]` in levels, then the ith element of the result is set to `NA`.

To obtain the integer values that are used for the encoding, use either `as.integer` or `unclass`. If the levels of the factor are themselves numeric, and you want to revert to the original numeric values (which do not need to correspond to the codes), the use of `as.numeric(levels(f))[f]` is recommended.

Great caution should be used when comparing factors since the interpretation depends on both the codes and the `levels` attribute. One should only compare factors that have the same sets of levels, in the same order. One scenario where comparison might be reasonable is to compare values between two different subsets of a larger data set, but here still caution is needed. You should ensure that unused levels are not dropped, as this will invalidate any automatic comparisons.

There are two tasks that are often performed on factors. One is to drop unused levels; this can be achieved by a call to `factor` since `factor(y)` will drop any unused levels from `y` if `y` is a factor. The second task is to coarsen the levels of a factor, that is, group two or more of them together into a single new level. The code below demonstrates one method for doing this.

```
> y = sample(letters[1:5], 20, rep = T)
> v = as.factor(y)
> xx = list(I = c("a", "e"), II = c("b", "c",
+       "d"))
> levels(v) = xx
> v

 [1] I  II II II I  I  II I  II I  I  II II I  II II II
[18] II II I
Levels: I II
```

Things are quite similar for ordered factors. They can be created by either by using the `ordered` argument to `factor` or with `ordered`.

Factors are instances of S3 classes. Ordinary factors have class *factor* and ordered factors have a class vector of length two with *ordered* as the additional element. An example of the use of an ordered factor is given below.

```
> z = ordered(y)
> class(z)

[1] "ordered" "factor"
```

Using a factor as an argument to the functions `matrix` or `array` coerces it to a character vector before creating the matrix.

2.2.4 Lists, environments and data frames

In this section we consider three different data structures that are designed to hold quite general objects. These data structures are sometimes called recursive since they can hold other R objects. The atomic vectors discussed above cannot.

There are actually two types of lists in R: `pairlists` and `lists`. We will not discuss pairlists in any detail. They exist mainly to support the internal code and workings of R. They are essentially lists as found in Lisp or Scheme (of the car, cdr, cons, variety) and are not particularly well adapted for use in most of the problems we will be addressing. Instead we concentrate on the `list` objects, which are somewhat more vector-like in their implementation and semantics.

2.2.4.1 Lists

Lists can be used to store items that are not all of the same type. The function `list` can be used to create a list. Lists are also referred to as generic vectors since they share many of the properties of vectors, but the elements are allowed to have different types.

```
> y = list(a = 1, 17, b = 4:5, c = "a")
> y

$a
[1] 1

[[2]]
[1] 17
```

```
$b
[1]  4 5

$c
[1]  "a"

> names(y)

[1]  "a" ""   "b" "c"
```

Lists can be of any length, and the elements of a list can be named, or not. Any R object can be an element of a list, including another list, as is shown in the code below. We leave all discussion of subsetting and other operations to Section 2.5.

```
> l2 = list(mn = mean, var = var)
> l3 = list(l2, y)
```

Exercise 2.5
Create a list of length 4 and then add a `dim` *attribute to it. What happens?*

2.2.4.2   Data frames

A `data.frame` is a special kind of list. Data frames were created to provide a common structure for storing rectangular data sets and for passing them to different functions for modeling and visualization. In many cases a data set can be thought of as a rectangular structure with rows corresponding to cases and columns corresponding to the different variables that were measured on each of the cases. One might think that a matrix would be the appropriate representation, but that is only true if all of the variables are of the same type, and this is seldom the case. For example, one might have height in centimeters, city of residence, gender and so on. When constructing the data frame, the default behavior is to transform character input into factors. This behavior can be controlled using the option `stringsAsFactors`.

Data frames deal with this situation. They are essentially a list of vectors, with one vector for each variable. It is an error if the vectors are not all of the same length. Data frames can often be treated like matrices, but this is not always true, and some operations are more efficient on data frames while others are less efficient.

Exercise 2.6
Look up the help page for `data.frame` *and use the example code to create a small data frame.*

2.2.4.3 Environments

An environment is a set of symbol-value pairs, where the value can be any R object, and hence they are much like lists. Originally environments were used for R's internal evaluation model. They have slowly been exposed as an R version of a hash table, or an associative array. The internal implementation is in fact that of a hash table. The symbol is used to compute the hash index, and the hash index is used to retrieve the value. In the code below, we create an environment, create the symbol value pair that relates the symbol a to the value 10 and then list the contents of the hash table.

```
> e1 = new.env(hash = TRUE)
> e1$a = 10
> ls(e1)

[1] "a"

> e1[["a"]]

[1] 10
```

Environments are different from lists in two important ways, and we will return to this point later in Section 2.5. First, for environments, the values can only be accessed by name; there is no notion of linear order in the hash table. Second, environments, and their contents, are not copied when passed as arguments to a function. Hence they provide one mechanism for pass-by-reference semantics for function arguments, but if used for that one should be cautious of the potential for problems. Perhaps one of the greatest advantages of the pass-by-value semantics for function calls is that in that paradigm function calls are essentially atomic operations. A failure, or error, part way through a function call cannot corrupt the inputs, but when an environment is used, any error part way through a function evaluation could corrupt the inputs.

The elements of an environment can be accessed using either the dollar operator, $, or the double square bracket operator. The name of the value desired must be supplied, and unlike lists partial matching is not used. In order to retrieve multiple values simultaneously from an environment, the mget function should be used.

In many ways environments are special. And as noted above they are not copied when used in function calls. This has at times surprised some users and here we give a simple example that demonstrates that these semantics mean that attributes really cannot be used on environments. In the code below, when e2 is assigned, no copy is made, so both e1 and e2 point to the same internal object. When e2 changes the attribute, it is changed for e1 as well. This is not what happens for most other types.

```
> e1 = new.env()
> attr(e1, "foo") = 10
> e1

<environment: 0x2b50b64>
attr(,"foo")
[1] 10

> e2 = e1
> attr(e2, "foo") = 20
> e1

<environment: 0x2b50b64>
attr(,"foo")
[1] 20
```

In the next code segment, an environment, e1, is created and has some values assigned in to it. Then a function is defined and that function has some free variables (variables that are not parameters and are not defined in the function). We then make e1 be the environment associated with the function and then the free variables will obtain values from e1. Then we change the value of one of the free variables by accessing e1 and that changes the behavior of the function, which demonstrates that no copy of e1 was made.

```
> e1 = new.env()
> e1$z = 10
> f = function(x) {
+      x + z
+ }
> environment(f) = e1
> f(10)

[1] 20

> e1$z = 20
> f(10)

[1] 30
```

Next, we demonstrate the semantics of rm in this context. If we remove e1, what should happen to f? If the effect of the command environment(f) = e1 was to make a copy of e1, then rm(e1) should have no effect, but we know that no copy was made and yet, as we see, removing `e1` appears to have no effect.

```
> rm(e1)
> f(10)

[1] 30

> f

function (x)
{
    x + z
}
<environment: 0x1f9f3b8>
```

What `rm(e1)` does is to remove the binding between the symbol `e1` and the internal data structure that contains the data, but that internal data structure is itself left alone. Since it can also be reached as the environment of `f`, it will remain available.

2.3   Managing your R session

The capabilities and properties of the computer that R is running on can be obtained from a number of builtin variables and functions. The variable `R.version$platform` is the canonical name of the platform that R was compiled on. The function `Sys.info` provides similar information. The variable `.Platform` has information such as the file separator. The function `capabilities` indicates whether specific optional features have been compiled in, such as whether jpeg graphics can be produced, or whether memory profiling (see Chapter 9) has been enabled.

```
> capabilities()

    jpeg      png    tcltk      X11     aqua http/ftp
    TRUE     TRUE     TRUE     TRUE     TRUE     TRUE
 sockets   libxml     fifo   cledit    iconv      NLS
    TRUE     TRUE     TRUE    FALSE     TRUE     TRUE
 profmem    cairo
    TRUE    FALSE
```

A typical session using R involves starting R, loading packages that will provide the necessary tools to perform the analysis you intend and then loading data, and manipulating that data in a variety of ways. For every R session you have a workspace (often referred to as the global environment) where any variables you create will be stored. As an analysis proceeds, it is often essential that you are able to manage your session and see what packages are attached, what variables you have created and often inspect them in some way to find an object you previously created, or to remove large objects that you no longer require.

You can find out what packages are on the search path using the `search` function and much more detailed information can be found using `sessionInfo`. In the code below, we load a Bioconductor package and then examine the search path. We use `ls` to list the contents of our workspace, and finally use `ls` to look at the objects that are stored in the package that is in position 2 on the search path. `objects` is a synonym for `ls` and both have an argument `all.names` that can be used to list all objects; by default, those that begin with a period are not shown.

```
> library("geneplotter")
> search()

 [1] ".GlobalEnv"            "package:geneplotter"
 [3] "package:annotate"      "package:xtable"
 [5] "package:AnnotationDbi" "package:RSQLite"
 [7] "package:DBI"           "package:lattice"
 [9] "package:Biobase"       "package:tools"
[11] "package:stats"         "package:graphics"
[13] "package:grDevices"     "package:utils"
[15] "package:datasets"      "package:methods"
[17] "Autoloads"             "package:base"

> ls(2)

 [1] "GetColor"              "Makesense"
 [3] "alongChrom"            "cColor"
 [5] "cPlot"                 "cScale"
 [7] "closeHtmlPage"         "dChip.colors"
 [9] "densCols"              "greenred.colors"
[11] "histStack"             "imageMap"
[13] "make.chromOrd"         "multidensity"
[15] "multiecdf"             "openHtmlPage"
[17] "panel.smoothScatter"   "plotChr"
[19] "plotExpressionGraph"   "saveeps"
[21] "savepdf"               "savepng"
[23] "savetiff"              "smoothScatter"
```

Most of the objects on the search path are packages, and they have the prefix `package`, but there are also a few special objects. One of these is `.GlobalEnv`, the *global environment*. As noted previously, environments are bindings of symbols and values.

Exercise 2.7
What does `sessionInfo` report? How do you interpret it?

2.3.1   Finding out more about an object

Sometimes it will be helpful to find out about an object. Obvious functions to try are `class` and `typeof`. But many find that both `str` and `object.size` are more useful.

```
> class(cars)

[1] "data.frame"

> typeof(cars)

[1] "list"

> str(cars)

'data.frame':         50 obs. of  2 variables:
 $ speed: num  4 4 7 7 8 9 10 10 10 11 ...
 $ dist : num  2 10 4 22 16 10 18 26 34 17 ...

> object.size(cars)

[1] 1248
```

The functions `head` and `tail` are convenience functions that list the first few, or last few, rows of a matrix.

```
> head(cars)

  speed dist
1     4    2
2     4   10
3     7    4
4     7   22
5     8   16
6     9   10
```

```
> tail(cars)

   speed dist
45    23   54
46    24   70
47    24   92
48    24   93
49    24  120
50    25   85
```

2.4 Language basics

Programming in R is carried out, primarily, by manipulating and modifying data structures. These different transformations and calculations are carried out using functions and operators. In R, virtually every operation is a function call and though we separate our discussion into operators and function calls, the distinction is not strong and the two concepts are very similar. The R evaluator and many functions are written in C but most R functions are written in R itself.

The code for functions can be viewed, and in most cases modified, if so desired. In the code below we show the code for the function colSums. To view the code for any function you simply need to type its name and the prompt and the function will be displayed. Functions can be edited using fix.

```
> colSums

function (x, na.rm = FALSE, dims = 1)
{
    if (is.data.frame(x))
        x <- as.matrix(x)
    if (!is.array(x) || length(dn <- dim(x)) < 2)
        stop("'x' must be an array of at least two dimensions")
    if (dims < 1 || dims > length(dn) - 1)
        stop("invalid 'dims'")
    n <- prod(dn[1:dims])
    dn <- dn[-(1:dims)]
    z <- if (is.complex(x))
        .Internal(colSums(Re(x), n, prod(dn), na.rm)) + (0+1i) *
            .Internal(colSums(Im(x), n, prod(dn), na.rm))
    else .Internal(colSums(x, n, prod(dn), na.rm))
```

```
    if (length(dn) > 1) {
        dim(z) <- dn
        dimnames(z) <- dimnames(x)[-(1:dims)]
    }
    else names(z) <- dimnames(x)[[dims + 1]]
    z
}
<environment: namespace:base>
```

Some functions cannot be accessed in this manner because the evaluator will attempt to parse them. For such functions you can use `get` to retrieve the function definition. The arithmetic operators fall into this category and next we show how to retrieve the definition for addition, +.

```
> get("+")

function (e1, e2)  .Primitive("+")
```

The body of this function is quite simple; it consists of one line, a call to `.Primitive`. And if you examine the body of `colSums`, you will notice that after some preliminaries there is a call to `.Internal`. These two functions, `.Primitive` and `.Internal`, provide fairly direct links between R level code and the internal C code that R is written in. Some functions are primitives for efficiency reasons and others for historical reasons. For such functions, users wanting to study the underlying code must explore the relevant C source code. Details and advice on such an investigation are provided in Appendix A of the R Extensions manual (R Development Core Team, 2007c).

Should you ever need to link C or other foreign language code to R, you will not use these functions; they are reserved for communications with the internal code for the R language. The appropriate user-level interfaces are discussed in Chapter 6.

2.4.1 Operators

In Table 2.1 we describe the different unary and binary operators in R. They are listed from highest precedence to lowest precedence. Operators on the same line are comma separated and have equal precedence. When operators of equal precedence occur in an expression, they are evaluated from left to right. This list can also be obtained from the manual page for `Syntax`.

As noted above, operators are a special syntax for ordinary function calls but the syntax of an operator is often more appealing. In the code below, we demonstrate this using the addition operator.

Operator	Description
[, [[	subscripting and subsetting
::, :::	name space access
$, @	access named components, access slots
^	exponentiation
+, -	unary plus and minus
:	sequence operator
%any%	special operators
*, /	multiply and divide
+, -	binary addition and subtract
< > <=, >=, == !=	comparisons
!	negation
&, &&	and
|, ||	or
~	define formulae
= , = >	right assignment
=	left assignment
= , < =	left assignment
?	help (both unary and binary)

Table 2.1: R operators, listed in precedence order.

```
> x = 1:4
> x + 5

[1] 6 7 8 9

> myP = get("+")
> myP

function (e1, e2)  .Primitive("+")

> myP(x, 5)

[1] 6 7 8 9
```

One class of operators of some interest is the set of operators of the form %any%. Some of these, such as %*%, are part of R but users can define their own using any text string in place of **any**. The function should be a function of two arguments, although currently this is not checked. In the example below we define a simple operator that pastes together its two arguments.

```
> "%p%" = function(x, y) paste(x, y, sep = "")
> "hi" %p% "there"

[1] "hithere"
```

2.5 Subscripting and subsetting

The S language has its roots in the Algol family of languages and has adopted some of the general vector subsetting and subscripting techniques that were available in languages such as APL. This is perhaps the one area where programmers more familiar with other languages fail to make appropriate use of the available functionality. Spending a few hours to completely familiarize yourself with the power of the subsetting functionality will be rewarded by code that runs faster and is easier to read.

There are slight differences between subsetting of vectors, arrays, lists, data.frames and environments that can sometimes catch the unwary. But there are also many commonalities. One thing to keep in mind is that the effect of NA will depend on the type of NA that is being used.

Subsetting is carried out by three different operators: the single square bracket [, the double square bracket [[, and the dollar, $. We note that each of these three operators are actually generic functions and users can write methods that extend and override them, see Chapter 3 for more details on object-oriented programming.

One way of describing the behavior of the single bracket operator is that the type of the return value matches the type of the value it is applied to. Thus, a single bracket subset of a list is itself a list. Thesingle bracket operator can be used to extract any number of values. Both [[and $ extract a single value. There are some differences between these two; $ does not evaluate its second argument while [[does, and hence one can use expressions. The $ operator uses partial matching when extracting named elements but [and [[do not.

```
> myl = list(a1 = 10, b = 20, c = 30)
> myl[c(2, 3)]

$b
[1] 20

$c
[1] 30
```

```
> myl$a

[1] 10

> myl["a"]

$<NA>
NULL

> f = "b"
> myl[[f]]

[1] 20

> myl$f

NULL
```

Notice that the first subsetting operation does indeed return a list, then that the $ subscript uses partial matching (since there is no element of myl named a) and that [does not. Finally, we showed that [[evaluates its second argument and $ does not.

2.5.1 Vector and matrix subsetting

Subsetting plays two roles in the S language. One is an extraction role, where a subset of a vector is identified by a set of supplied indices and the resulting subset is returned as a value. Venables and Ripley (2000) refer to this as *indexing*. The second purpose is subset assignment, where the goal is to identify a subset of values that should have their values changed; we call this subset assignment.

There are four basic types of subscript indices: positive integers, negative integers, logical vectors and character vectors. These four types cannot be mixed; only one type may be used in any one subscript vector. For matrix and array subscripting, one can use different types of subscripts for the different dimensions. Not all vectors, or recursive objects, support all types of subscripting indices. For example, atomic vectors cannot be subscripted using $, while environments cannot be subscripted using [. Missing values can appear in the index vector and generally cause a missing value to appear in the output.

2.5.1.0.1 Subsetting with positive indices Perhaps the most common form of subsetting is with positive indices. Typically, a vector containing the integer subscripts corresponding to the desired values is used. Thus, to extract entries one, three and five from a vector, one can use the approach demonstrated in the next code chunk.

```
> x = 11:20
> x[c(1, 3, 5)]

[1] 11 13 15
```

The general rules for subsetting with positive indices are:

- A subscript consisting of a vector of positive integer values is taken to indicate a set of indexes to be extracted.

- A subscript that is larger than the length of the vector being subsetted produces an NA in the returned value.

- Subscripts that are zero are ignored and produce no corresponding values in the result.

- Subscripts that are NA produce an NA in the result.

- If the subscript vector is of length zero, then so is the result.

Some of these rules are demonstrated next.

```
> x = 1:10
> x[1:3]

[1] 1 2 3

> x[9:11]

[1]  9 10 NA

> x[0:1]

[1] 1

> x[c(1, 2, NA)]

[1]  1  2 NA
```

Exercise 2.8
Use the seq *function to generate a subscript vector that selects those elements of a vector that have even-numbered subscripts.*

2.5.1.0.2 Subsetting with negative indices In many cases it is simpler to describe the values that are *not* wanted, than to specify those that are. In this case, users can use negative subscript indices; the general rules are listed below.

- A subscript consisting of a vector of negative integer values is taken to indicate the indexes that are not to be extracted.

- Subscripts that are zero are ignored and produce no corresponding values in the result.

- `NA` subscripts are not allowed.

- A zero length subscript vector produces a zero length answer.

- Positive and negative subscripts cannot be mixed.

Exercise 2.9
Use the function `seq` to generate a sequence of indices so that those elements of a vector with odd-numbered indices can be excluded. Verify this on the built-in `letters` data. Verify the statement about zero length subscript vectors.

2.5.1.0.3 Subsetting with character indices Character indices can be used to extract elements of named vectors, lists. While technically having a `names` attribute is not necessary, the only possible result if the vector has no names is `NA`. There is no way to raise an error or warning with character subscripting of vectors or lists; for vectors `NA` is returned and for lists `NULL` is returned. Subsetting of matrices and arrays with character indices is a bit different and is discussed in more detail below.

For named vectors, those elements whose names match one of the names in the subscript are returned. If names are duplicated, then only the value corresponding to the first one is returned. `NA` is returned for elements of the subscript vector that do not match any name. A character `NA` subscript returns an `NA`.

If the vector has duplicated names that match a subscript, only the value with the lowest index is returned. One way to extract all elements with the same name is to use `%in%` to find all occurrences and then subset by position, as demonstrated in the example below.

```
> x = 1:5
> names(x) = letters[1:5]
> x[c("a", "d")]

a d
1 4
```

```
> names(x)[3] = "a"
> x["a"]

a
1

> x[c("a", "a")]

a a
1 1

> names(x) %in% "a"

[1]  TRUE FALSE  TRUE FALSE FALSE
```

Exercise 2.10
Verify that vectors can have duplicated names and that if a subscript matches a duplicated name, only the first value is returned. What happens with x[NA], and why does that not contradict the claims made here about NA subscripts? Hint: it might help to look back at Section 2.1.4.

Lists subscripted by NA, or where the character supplied does not correspond to the name of any element of the list, return NULL.

2.5.1.0.4 Subsetting with logical indices A logical vector can also be used to subset a vector. Those elements of the vector that correspond to TRUE values in the subscript vector are selected, those that correspond to FALSE values are excluded and those that correspond to NA values are NA. The subscript vector is repeated as many times as necessary and no warning is given if the length of the vector being subscripted is not a multiple of the subscript vector.

If the subscript vector is longer than the target, then any entries in the subscript vector that are TRUE or NA generate an NA in the output.

```
> (letters[1:10])[c(TRUE, FALSE, NA)]

[1] "a" NA  "d" NA  "g" NA  "j"

> (1:5)[rep(NA, 6)]

[1] NA NA NA NA NA NA
```

Exercise 2.11
Use logical subscripts to extract the even-numbered elements of the `letters`
vector.

2.5.1.0.5 Matrix and array subscripts Empty subscripts are most often used for matrix or array subsetting. An empty subscript in any dimension
indicates that all entries in that dimension should be selected. We note that
`x[]` is valid syntax regardless of whether `x` is a list, a matrix, an array or a
vector.

```
> x = matrix(1:9, nc = 3)
> x[, 1]

[1] 1 2 3

> x[1, ]

[1] 1 4 7
```

One of the peculiarities of matrix and array subscripting is that if the resulting value is such that the result has only one dimension of length larger
than one, and hence is a vector, then the dimension attribute is dropped and
the result is returned as a vector. This behavior often causes hard-to-find and
hard-to-diagnose bugs. It can be avoided by the use of the `drop` argument to
the subscript function, `[`. Its use is demonstrated in the code below.

```
> x[, 1, drop = FALSE]

     [,1]
[1,]    1
[2,]    2
[3,]    3

> x[1, , drop = FALSE]

     [,1] [,2] [,3]
[1,]    1    4    7
```

Since arrays and matrices can be treated as vectors, and indeed that is how
they are stored, it is important to know the relationship between the vector
indices and the array indices. Arrays and matrices in S are stored in column
major order. This is the form of storage used by FORTRAN and not that
used by C. Thus the first, or left-most, index moves the fastest and the last,

or right-most, index the slowest, so that a matrix is filled column by column (the row index changes). This is often referred to as being in *column major* order. The function `matrix` has an option named `byrow` that allows the matrix to be filled row by row, rather than column by column.

Exercise 2.12
Let x be a vector of length 10 and has a dimension attribute so that it is a matrix with 2 columns and 5 rows. What is the matrix location of the 7th element of x? That is, which row and column is it in? Alternatively, which element of x is in the second row, first column?

Finally, an array may be indexed by a matrix. If the array has k dimensions, then the matrix must be of dimension l by k and must contain integers in the range 1 to k. Each row of the index array is interpreted as identifying a single element of the array. Thus the subscripting operation returns l values. A simple example is given below. If the matrix of subscripts is either a character matrix or a matrix of logical values, then it is treated as if it were a vector and the dimensioning information is ignored.

```
> x = array(1:27, dim = c(3, 3, 3))
> y = matrix(c(1, 2, 3, 2, 2, 2, 3, 2, 1), byrow = TRUE,
+       ncol = 3)
> x[y]

[1] 22 14  6
```

Character subscripting of matrices is carried out on the row and column names, if present. It is an error to use character subscripts if the row and column names are not present. Attaching a `dim` attribute to a vector removes the `names` attribute if there was one. If a `dimnames` attribute is present, but one or more of the supplied character subscripts is not present, a subscript out of bounds error is signaled, which is quite different from the way vectors are treated. Arrays are treated similarly, but with respect to the names on each of the dimensions.

For `data.frames` the effects are different. Any character subscript for a row that is not a row name returns a vector of NAs. Any subscript of a column with a name that is not a column name raises and error.

Exercise 2.13
Verify the claims made for character subsetting of matrices and data.frames.

Arrays and matrices can always be subscripted singly, in which case they are treated as vectors and the dimension information is disregarded (as are the `dimnames`). Analogously, if a `data.frame` is subscripted with a single subscript, it is interpreted as list subscripting and the appropriate column is selected.

2.5.1.0.6 **Subset assignments** Subset expressions can appear on the left side of an assignment. If the subset is specified using positive indices, then the given subset is assigned the values on the right, recycling the values if necessary. Zero subscripts and NA subscripts are ignored.

```
> x[1:3] = 10
> x

, , 1

     [,1] [,2] [,3]
[1,]   10    4    7
[2,]   10    5    8
[3,]   10    6    9

, , 2

     [,1] [,2] [,3]
[1,]   10   13   16
[2,]   11   14   17
[3,]   12   15   18

, , 3

     [,1] [,2] [,3]
[1,]   19   22   25
[2,]   20   23   26
[3,]   21   24   27
```

Negative subscripts can appear on the the left side of an assignment. In this case the given subset is assigned the values on the right side of the assignment, recycling the values if necessary. Zero subscripts are ignored and NA subscripts are not permitted.

```
> x = 1:10
> x[-(2:4)] = 10
> x

[1] 10  2  3  4 10 10 10 10 10 10
```

For character subscripts, the selected subset is assigned the values from the right side of the assignment, recycling if necessary. Missing values (character

NA) create a new element of the vector, even if there is already an element with the name NA. Note that this is quite different from the effect of a logical NA, which has no effect on the vector.

Exercise 2.14
Verify the claims made about negative subscript assignments. Create a named vector, x, and set one of the names to NA. What happens if you execute x[NA]=20 *and why does that not contradict the statements made above? What happens if you use* x[as.character(NA)]=20?

In some cases leaving all dimensions out can be useful. For example, x[], selects all elements of the vector x and it does not change any attributes.

```
> x = matrix(1:10, nc = 2)
> x[] = sort(x)
```

2.5.1.0.7   Subsetting factors   There is a special method for the single bracket subscript operator on factors. For this method the `drop` argument indicates whether or not any unused levels should be dropped from the return value. The [[operator can be applied to factors and returns a factor of length one containing the selected element.

2.6   Vectorized computations

By vectorized computations we mean any computation, by the application of a function call, or an operator (such as addition), that when applied to a vector automatically operates directly on all elements of the vector. For example, in the code below, we add 3 to all elements of a simple vector.

```
> x = 11:15
> x + 3

[1] 14 15 16 17 18
```

There was no need to make use of a `for` loop to iterate over the elements of x. Many R functions and most R operators are vectorized. In the code below `nchar` is invoked with a vector of months names and the result is a vector with the number of characters in the name of each month; there is no need for an explicit loop.

```
> nchar(month.name)

[1] 7 8 5 5 3 4 4 6 9 7 8 8
```

2.6.1 The recycling rule

Since vectorized computations occur on most arithmetic operators we may encounter the problem of adding together two vectors of different lengths and some rule is needed to describe what will happen in that situation. This is often referred to as the recycling rule.

The recycling rule states that the shorter vector is replicated enough times so that the result has at least the length of the longer vector and then the operator is applied to the resulting elements, of that modified vector, that correspond to elements of the longer vector. If the shorter vector is not an even multiple of the longer vector, a warning will be signaled.

```
> 1:10 + 1:3

[1]  2  4  6  5  7  9  8 10 12 11
```

When a binary operation is applied to two or more matrices and arrays, they must be conformable, which means that they must have the same dimensions. If only one operand is a matrix or array and the other operands are vectors, then the matrix or array is treated as a vector. The result has the dimensions of the matrix or array. Dimension names, and other attributes, are transferred to the output.

In general the attributes of the longest element, when considered as a vector, are retained for the output. If two elements of the operation are of the same length, then the attributes of the first one, when the statement is parsed left to right, are retained.

Any vector or array computation where the recycling rule applies and one of the elements has length zero returns a zero length result. Some examples of these behaviors are given below.

```
> 1:3 + numeric()

numeric(0)

> 1:3 + NULL

numeric(0)
```

```
> x = matrix(1:10, nc = 2)
> x + (1:2)

     [,1] [,2]
[1,]    2    8
[2,]    4    8
[3,]    4   10
[4,]    6   10
[5,]    6   12
```

2.7   Replacement functions

In R it is sometimes helpful to appear to directly change a variable. We saw some examples of this with subassignment; e.g., x[2] = 10 gives the impression of having changed the value of x. Similarly, changing the names on a vector can be handled using names(x) = newVal. Some reflection on the fact that R is a pass-by-value language and that all operations are function calls means that, in principle, such an operation is not possible. That is, it is not possible to change x, since the function operates on a copy of x.

Following Venables and Ripley (2000), any assignment where the left-hand side is not a simple identifier will be described as a replacement function. These functions achieve their objective by rewriting the call in such a way that the named variable (x in the examples above) is explicitly reassigned.

We show how these two commands would be rewritten, below. You can make these calls directly, if you wish, and all functions are documented and can be inspected, just like any other functions.

```
> x = 1:4
> x = `[<-`(x, 2, value = 10)
> x

[1]  1 10  3  4
```

And for the names example:

```
> names(x) = letters[1:4]
> names(x)

[1] "a" "b" "c" "d"
```

```
> x = "names<-"(x, LETTERS[1:4])
> x

A  B  C  D
1 10  3  4
```

Without the explicit assignment, the value of x cannot be changed because
the functions [<- and **names<-** work on copies of their arguments.

To write a replacement function, you must make sure that the last two
characters of the name are <-, and usually that means you will need to enclose
the name in double quotes to prevent the evaluator from attempting to carry
out an explicit assignment. You must also make sure that the return value is
the modified copy of the object you wanted to modify. R will automatically
rewrite the code to carry out an assignment, so the return value must be
appropriate for that operation. If it is not, no error will be signaled but the
result will probably not be useful.

The last argument to a replacement function must be named **value** and it
will be matched to the value of the right-hand side of the assignment.

Exercise 2.15
Write a replacement function, called rowrep, *that replaces the indicated row*
of a matrix, x *say, with the value on the right-hand side of the assignment.*
That is, we want a syntax like rowrep(x, 4) = c(11, 12) *to replace the fourth*
row of x *with the values 11 and 12.*

2.8   Functional programming

In many ways R can be considered as a functional programming language.
Functions can be passed as arguments to functions, and returned as values.
But other aspects of functional programming are also available in R. The functions are written in R, and hence do not provide performance improvements
over performing the computations from first principles, but that could change.
They do however provide useful abstractions and can yield code that is easier
to comprehend. It is instructive to look at the implementations.

There are four functions that are part of R that implement some of the
ideas of functional programming.

Reduce Reduce takes a function of two arguments and an input vector, and
successively combines the elements of that vector.

Filter Filter extracts the elements of a vector conditional on a logical function, returning true when applied to that element.

Map Map applies a function to the corresponding elements of an arbitrary number of input vectors.

Negate Negate creates a negation of a given function.

Map is a lot like the `apply` family of functions (Section 8.2.2), and if named arguments are given, where the names match the name of a formal argument to the function being mapped, it is used.

```
> Map(paste, 1:4, letters[1:4], sep = "_")

[[1]]
[1] "1_a"

[[2]]
[1] "2_b"

[[3]]
[1] "3_c"

[[4]]
[1] "4_d"
```

Filter is also similar to the `apply` family of functions, with the distinction that it filters out values that fail the predicate. A standard idiom for removing missing values from a vector is to find them, and then remove them via subsetting. In the code below, we first construct a vector with some missing values, and then demonstrate two ways of removing NAs.

```
> set.seed(123)
> x = rnorm(1000)
> x = ifelse(abs(x) > 2.2, NA, x)
> y = x[!is.na(x)]
> y2 = Filter(Negate(is.na), x)
> all(y == y2)

[1] TRUE
```

2.9   Writing functions

A very powerful aspect of the R language is the fact that it is easy to write your own functions and to make use of them. All functions take inputs, which are referred to as the arguments, and return a single value. In R the value returned by a function is either the value that is explicitly returned by a call to the function `return` or it is simply the value of the last expression.

A function is defined by using the keyword `function`, which is followed by an opening parenthesis, (, then a comma separated list of *formal arguments*, a closing), and then the expressions that form the *body* of the function. If there is a single expression, it can be entered directly and when there are several expressions, or statements that will be executed, they must be enclosed in braces (`{}`).

In the code below we define two simple functions to compute the square of their input values. In both cases there is a single formal argument, named x. Since both functions contain a single line, no braces were used, and in `sq1` there is an explicit call to `return`, while in the second case we rely on the fact that the value returned by the function is the value of the last statement executed.

```
> sq1 = function(x) return(x * x)
> sq2 = function(x) x * x
```

These functions are *vectorized*. Try evaluating both versions with inputs `1:10`. What values do you get?

Exercise 2.16
Write a function that takes a string as input and returns that string with a caret prepended. If you name your function `ppc` *then we want* `ppc("xx")` *to return* `"^xx"`. *You will probably find the function* `paste` *helpful.*

Many more details on writing functions are given in other texts, such as (Becker et al., 1988; Chambers, 1998; Venables and Ripley, 2000) and we will not repeat those details here. We briefly note that it is possible to set default values for the formal arguments and remind readers that there is a special formal argument, represented by three dots, that allows for an arbitrary number of arguments to be passed to a function. There are specific rules (described in the references given) for argument matching, etc. Partial matching is used, and when an ... argument appears as a formal argument, then any arguments that appear after it must be matched exactly.

An example

One of the tasks that often needs to be carried out is some form of standardizing of data. In the gene expression context we often want to subtract some notion of the center, and divide by some notion of the variability. Typically this is done on a gene by gene basis (and hence on the rows of the expression matrix).

In R, there is a function called `scale` that performs a specific type of centering and scaling but on the columns, rather than the rows (the reason for this is that the usual form of statistical data is to have the cases represented as rows; gene expression data are an exception to this, where the cases are the columns). There is another R function, `sweep`, that can be used to do more general forms of standardization.

Exercise 2.17
Write an R function that takes a matrix as input and standardizes each row by subtracting the median and dividing by the MAD. You might want to look at the functions `mad` and `median`. Next, generalize your function so that these are the default values, but allow the user to select other measures of the center and spread, should they want to use them.

2.10   Flow control

Carrying out a specific computation for each element of a vector or a list is one of the fundamental computational tasks that must be performed in any computer program. In some languages, iteration is the main tool that is used but in others recursion is more common. In R, both recursion and iteration can be used, and which will be more efficient depends to some extent on the data structures that are being used. In this section we primarily discuss iteration; recursion is presented later in this chapter. We also note that one of the primary tools for data processing is the `apply` family of functions. We defer discussion of these functions to Chapter 8.

In many other computer languages, explicit iteration is needed to operate on all of the elements of a vector, but in R it is generally more efficient to make use of vectorized computations (as described in Section 2.6).

There are three basic paradigms for iteration: `for`, `repeat` and `while`. These functions need to be quoted when invoking the help system to find out more about them. The syntax of these is given below.

```
for (var in seq) expr
while (cond) expr
repeat expr
```

In these constructs:

expr is any valid R expression, and is often a compound expression, which is a series of expressions contained with in curly braces.

cond is a length one logical value, although in recent versions of R, length one numeric values also work, where zero corresponds to FALSE and any non-zero value corresonds to TRUE.

var is a dummy variable, that takes values in seq and can be used within expr. In R, this variable remains after the for loop has finished and retains the last value it had.

seq is any vector or list.

Each of these three functions evaluates the expression expr repeatedly. The first, for, iterates through the values in seq; the latter two rely on the expr changing state, or the explicit use of break to break out of the iteration. In addition to these three functions, there are two special values, break and next, that can be used to control the iteration. The use of break halts the execution of the inner-most loop and passes control to the next statement. The use of next halts the execution of the current expr and begins the next evaluation. When using either repeat or while, some care is needed to avoid an infinite loop, that is, a loop that iterates without end.

The most common idiom used with the for loop is for(i in 1:n), but in such cases the use of for(i in seq_len(n)) is better. The form for(var in seq), where var assumes the value of each element of seq in turn, may be more efficient in some cases.

```
> for (i in 1:3) print(i)

[1] 1
[1] 2
[1] 3

> for (i in 1:5) if (i > 3) break
> i

[1] 4
```

Exercise 2.18

Use repeat, next and break to print the odd integers between 1 and 10. Repeat this exercise using while, instead of repeat, and print the even integers.

2.10.1   Conditionals

There are three functions that can be used for conditional evaluation: `if`, `ifelse` and `switch`. The syntax of the `if` statement is

```
if( cond ) expr1 else expr2
```

where `cond` and `expr` are as described previously.

The condition `cond` is evaluated, and if it evaluates to `TRUE`, then `expr1` is evaluated; if `cond` evaluates to `FALSE`, then `expr2` is evaluated. In recent versions of R, if `cond` evaluates to zero, then it is treated as `FALSE`; and if `cond` evaluates to any non-zero number, it is treated as if it were `TRUE`. If the length of `cond` is larger than one, a warning is signaled and only the first element is used in determining which expression to evaluate.

The `else` clause is optional. If the command is run at the command prompt and there is an `else` clause, then either all the expressions must be enclosed in curly braces or the `else` must be on the same line as `expr1`. The reason for this behavior is quite simple; since the `else` clause is optional and since R does not use end of statement syntax, the R evaluator must conclude that it has a syntactically valid statement (after parsing `expr1`) and it will evaluate it. When the same code is used inside of a function, there is no need for the `else` clause to be on the same line as `expr1`, since in this case the parser will have access to the whole function and can determine whether there is a following `else` clause or not.

The function `ifelse` takes three arguments, `test`, `yes` and `no`, and if needed replicates values for `yes` and `no` so that they are the same length as `test`. All elements of `test` that evaluate to `TRUE` are replaced by the corresponding value in `yes` and those values of `test` that evaluate to `FALSE` are replaced by the corresponding values in `no`. Additionally, if `test` has a dimension attribute, it is retained. All of the arguments are evaluated, so constructs such as `ifelse(x<0, 0, log(x))` will produce warnings since `log` is performed on all values of `x`.

```
> x = matrix(1:10, nc = 2)
> ifelse(x < 2, x, c(10, 11, 12))

     [,1] [,2]
[1,]    1   12
[2,]   11   10
[3,]   12   11
[4,]   10   12
[5,]   11   10
```

Exercise 2.19
Explain the output in the preceding code chunk.

We conclude the discussion of conditional evaluation with the function
`switch`. The first argument to `switch` is an expression to evaluate. Any number of additional arguments can be supplied, and they can be either named
or not. If the value of the expression is numeric, then the corresponding
additional argument is evaluated and returned. If the expression returns a
character value, then the additional argument with the matching name will
be evaluated and returned. If no argument has a matching name, then the
value of the first unnamed argument is returned.

Partial matching can be problematic when using `switch`. Since the first
argument is named `EXPR`, any named argument that partially matches this
could be inadvertently used as the expression. Always specifically naming the
`EXPR` is a good defensive strategy.

The code chunk below is taken from the manual page for the function `switch`
and exemplifies many of the uses this function is put to.

```
> centre = function(x, type) {
+     switch(type, mean = mean(x), median = median(x),
+         trimmed = mean(x, trim = 0.1))
+ }
> x = rcauchy(10)
> centre(x, "mean")

[1] 2.829261

> centre(x, "median")

[1] 0.4971483

> centre(x, "trimmed")

[1] 0.7449222
```

Exercise 2.20
*Many implementations of `switch` include a capability to return a default value
if there is no match. Show that this can be done in R using `switch` and named
arguments, but that for numbered arguments this is not possible.*

2.11 Exception handling

Exception handling is the process of dealing with the failure of a computation to complete successfully and in some cases to allow the user to interrupt

computations. R has a number of tools that allow for very general exception handling. Here we restrict our attention to some of the more basic uses and encourage interested readers to investigate the available resources for more details and examples if the coverage here is not sufficient.

The two most common sorts of exceptions are errors and warnings. Errors can be raised by a call to `stop` and warnings can be raised by a call to `warning`. The typical behavior for an error is to halt the current evaluation and return control to the top-level R prompt. However, in many situations this may not be desired. A common situation is where a large simulation is being run and the failure of one run should not halt the entire simulation. Control over error and warning handling is provided by the exception handling system.

The default behavior for warnings is to wait until the current evaluation is finished and to then print the warnings that occurred during the evaluation process. Users can control the behavior by making use of various R options. In particular, the options `warn`, `error` and `show.error.messages` can be used, but there are other options that control other aspects. Their values can be changed by calling the function `options` with the appropriate arguments. More detailed discussion of settings for the `error` option that allow for debugging and inspecting the evaluation process are provided in Chapter 9. The present discussion focuses on programming paradigms to ensure that evaluation proceeds as desired when an exception is signaled.

Perhaps the simplest interface to use is the function `try`, which takes two arguments, `expr` and `silent`. The first of these is an expression to evaluate. When invoked, `try` will evaluate the expression provided. If the expression evaluates without error, then the value is returned. Otherwise, an instance of the class *try-error* is returned. More details on the class system are provided in Chapter 3 but for now simply consider that a different value is returned if the evaluation of the provided expression causes an error. The built-in error handling system will cause the error to be printed at the console unless either the `show.error.messages` option has been set to `FALSE` or the second argument to `try` has been set to `TRUE`. The return value must be explicitly tested to see if it is a *try-error*. Recently, `try` was reimplemented using `tryCatch`, which is described below.

One example of its use is in the function `testBioCConnection` in the **Biobase** package. The goal is to see whether the user has access to the Internet and can find and successfully read from the Bioconductor repository. A small piece of the relevant code is given below.

```
> {   top = options(show.error.messages=FALSE)
+     test = try(readLines(biocURL)[1])
+     options(top)
+     if (inherits(test,"try-error"))
+         return(FALSE)
+     else
```

```
+       close(biocURL)
+    return(TRUE)
+ }
```

We see that first, the option that controls showing error messages is set to
FALSE, then there is a call to try, the option is restored to its previous value
and then we test to see if the attempt to read was successful.

There are a few interesting points here, among them the practice of assigning the return value from a call to options. This allows you to then reset that
value, regardless of its initial value. For this example, regardless of whether
show.error.messages was set to TRUE or FALSE on entry to the function, it will
have the save value on exit.

A second, and in some ways simpler, mechanism for conditionally evaluating expressions is provided by the tryCatch function; the first argument is
evaluated and if it returns without raising a condition, then its value is returned. If a condition is raised, or signaled due to the evaluation of the first
argument, then the other named arguments are searched to see if any one
of them has a name that corresponds to the class of the condition that was
raised. If a match is found, then the handler supplied is invoked.

In the code below we demonstrate the use of tryCatch on a simple example.
In the first two cases, two handlers are established: one for **errors** and the
other for **warnings**.

```
> foo = function(x) {
+    if (x < 3)
+        list() + x
+    else {
+        if (x < 10)
+            warning("ouch")
+        else 33
+    }
+ }
> tryCatch(foo(2), error = function(e) "an error",
+    warning = function(e) "a warning")

[1] "an error"

> tryCatch(foo(5), error = function(e) "an error",
+    warning = function(e) "a warning")

[1] "a warning"

> tryCatch(foo(29))

[1] 33
```

tryCatch has an additional argument named `finally`, which is an expression to be evaluated before returning and exiting. The `finally` expression is evaluated in the context in which the call `tryCatch` was made and hence none of the handlers are established.

Users can define new classes of conditions and these classes can be handled by `tryCatch` by setting appropriate handler functions. For example, in the code below we define a condition that is appropriate when a file is not found. When this signal is thrown, the user interface could invoke a file browser allowing the user to select the appropriate file interactively. In the code chunk below, we first present a function that can be used to create the appropriate conditions, then we create a condition that can be signaled. In the call to `tryCatch`, we signal the condition and it is then handled by one of the established handlers.

The order in which the character strings provided in the `class` attribute is important. We discuss the properties of the S3 object system in more detail in Chapter 3 but for now it is sufficient to know that the order is important. The handler that is invoked by `tryCatch` is determined by this order, the first one that matches an element of the class attribute of the condition is used.

```
> FNFcondition = function (message, call = NULL){
+     class = c("fileNotFound", "error", "condition")
+     structure(list(message = as.character(message),
+                    call = call), class = class)
+ }
> v1 = FNFcondition("file not found")
> tryCatch( signalCondition(v1), fileNotFound = function(e) e )

<fileNotFound: file not found>

> tryCatch( signalCondition(v1),
+           condition = function(e) "condition" )

[1] "condition"
```

Another important aspect of control of evaluation of a program is the correct handling of signals. Signals are a software mechanism that allows for the reporting of exceptional situations (out of memory, or invalid memory access) as well as reporting and detecting asynchronous events. They are often raised by the operating system but there are many ways processes can use them to communicate with other processes. When users want to interrupt a long running computation, or break out of an infinite loop, they commonly attempt to use an interrupt (ctrl-C), which is a signal that is raised when the user hits the control and C keys simultaneously. In the code below, the expression `repeat(readline())` results in an infinite loop, which can be broken out of by sending a user interrupt (ctrl-C), which is caught by the `interrupt` handler defined in the call to `tryCatch`.

```
> tryCatch(repeat(readline()),
+      interrupt=function(e) print("howdy"))
```

Restarts are another part of the condition handling system in R. Making use of the restart requires that the user set an error handler that will enter the browser, or more generally, a calling handler that looks at the available restarts. The example below is adapted from a talk by L. Tierney in 2003. The function establishes a restart and then attempts to download a file, and if it fails, then the restart is available.

```
> downloadWithRestarts = function(url, destfile, ...){
+     repeat
+         withRestarts(return(download.file(url, destfile, ...)),
+                      retryDownload = function() NULL,
+                      tryNewUrl = function(newUrl)
+                      url <<- newUrl)
+ }
```

If we then establish this within a call to `withCallingHandlers`, then if an error occurs, the user will be placed into the `browser` and can call `invokeRestart` to access one of the two restarts that were established; namely, `retryDownload` and `tryNewUrl`.

```
> withCallingHandlers(downloadWithRestarts("http://foo.bar.org",
+     "xyz"), error = function(e) {
+     cat("Error:", conditionMessage(e), "\n")
+     browser()
+ })
```

Since the URL does not exist, the result is to signal an error, which then invokes the `browser` and the user can then call the `retryDownload` restart to try to download again, or the `tryNewUrl` restart to supply a new URL to try.

The code below is based on an example posted to the R help mailing list. It demonstrates another use of `withCallingHandlers`. If an error is signaled, it is caught by the `error` handler that invokes the `skipError` restart, which returns `NULL` and the `for` loop continues.

```
> even = function(i) i %% 2 == 0
> testEven = function(i) if (even(i) ) i else stop("not even")
> vals = NULL
> withCallingHandlers({
+      for (i in seq_len(10)) {
+          val = withRestarts(testEven(i),
+          skipError=function() return(NULL))
+          if (!is.null(val))
+                vals = c(vals, val)
+      }},
+          error=function(e) invokeRestart("skipError"))
> vals

[1]  2  4  6  8 10
```

2.12   Evaluation

Computer languages provide a level of abstraction that allows programmers, and users, the ability to express certain ideas symbolically. In many cases the symbolic statements constitute a program, or a function that may be evaluated many times, with potentially different inputs. In other cases, with interactive languages such as R, users will type statements directly in the evaluator and immediately see the consequences of their commands. In this section we consider the process of evaluating statements in R. Since evaluation is a major component of computer use, it is essential that programmers both understand the model that is used in any language and that they be able to control the evaluation process to reliably get the intended answers.

R is a pass-by-value language. That means that when a function call is made, the arguments are copied. For efficiency reasons the actual implementation attempts to only duplicate if any modification is made to the supplied argument and so is sometimes described as *copy on modify*.

Consider the expression: 1+10. In this expression there are three distinct symbols, 1, + and 10. The first and third we think of as representing values: the numbers one and ten, respectively. But what about +? Perhaps we have seen this so much that we feel that it too is really a value – but in reality it is merely a symbol, and the correct value, which may be a function, must be obtained.

While we will not get into the explicit detail of how the parser works, we note that many of the details are described in Writing R Extensions (R

Development Core Team, 2007c). The expression 1+10 is parsed into a call
to the function + with two arguments: 1 and 10. And the result is, of course,
11. The main point of this example is to draw attention to the fact that
everything is a symbol and that all symbols must be resolved to obtain values.
Some values are functions that will be invoked, while others are values that
are returned and printed.

2.12.1 Standard evaluation

The rules that R uses for associating values with symbols are relatively
straightforward. We begin by describing what happens on interactive use,
and follow that with a description of the evaluation process during function
invocation. So we return to our simple example of evaluating 1+10.

The search path plays an important role in the evaluation process (see
Section 2.3 for more details). When the R evaluator is looking for a value to
associate with the symbol +, it traverses the search path from the first element
to the last. In most cases the first instance of a symbol + that is encountered
will be used. However, when R knows the type of the object it is looking for,
in this case a function, it will skip over bindings to values that are not of the
correct type. We can determine which version of a symbol will be used by
using the function find, and we can examine the associated value using get.
If other elements of the search path contain a binding for the symbol being
sought, they are said to be *masked* by the first definition.

The search path is intrinsically dynamic in nature. Function calls can have
the side effect of attaching new packages and hence of altering the bindings
that will be used. Packages can also be detached and function definitions can
be made in the global environment at virtually any time. Programmers are
cautioned not to rely on the order in which packages are attached; explicit
use of name spaces and lexical scope should be preferred.

```
> find("+")

[1] "package:base"

> get("+")

function (e1, e2)  .Primitive("+")
```

Users are free to override system functions, at their own risk. In the next
code chunk we first assign a new value for + and then evaluate some example
code and then remove the value we have assigned, since we do not want to
interfere with the usual definition of +.

```
> assign("+", function(e1, e2) print("howdy"))
> 1 + 10

[1] "howdy"

> rm("+")
> 1 + 10

[1] 11
```

The binding of the symbol + occurs in the .GlobalEnv, and so removing it will restore the standard system function.

The function find will find all instances of a symbol on the search path, and has arguments that allow you to specify the type of object that you are searching for. The one that is closest to the start of the search path (nearest to position 1) will be used, under normal conditions. The other bindings are said to be *masked*. This process is essentially the same for all function lookups, including those defined in system functions. This has some potential for problems and for naming conflicts between unrelated packages. *Name spaces* provide a mechanism that allows programmers more control over which bindings will be used, and we review those details in Section 2.12.7.

2.12.2   Non-standard evaluation

In R, there are a number of functions that do not use the standard evaluation rules. The use of non-standard evaluation is discouraged, and users should only make use of it if there are compelling reasons. Some functions have retained their behavior for compatibility reasons. You should avoid using non-standard rules. Readers interesed in this topic should consult T. Lumley's document developer.r-project.org/nonstandard-eval.pdf, which details the current non-standard evaluation paradigms and some of the pitfalls that should be considered. Here, we very briefly show one method that is used in some functions, mostly so that the reader is aware of how this code works, and not to encourage the use of this paradigm.

Perhaps the most common non-standard situation is where a function does not evaluate its argument, but rather makes use of the name provided when the function is invoked. The misnamed library function does just this. You should have wondered how the following code works.

```
> library(tools)
```

Since tools is not a symbol, and it is not a character literal (enclosed in quotes), then this evaluation should, under standard evaluation rules, fail since the first thing that should happen is for the arguments to be evaluated. Instead, the following code, extracted from `library`,

```
if (!character.only)
    package = as.character(substitute(package))
```

shows how standard evaluation is bypassed. This is a fairly standard paradigm in R. The use of `substitute` on a formal argument to a function retrieves the actual symbol that was supplied. And `substitute` is documented to not evaluate its arguments, so in this case the formal argument `package` is never evaluated.

Non-standard evaluation also arises quite often in functions that fit various models to data, such as `lm` and `glm`.

2.12.3   Function evaluation

Function evaluation is essentially the same as that of evaluation of statements typed directly to the evaluator. The major differences arise due to the fact that when one function is invoked, it is very likely that there are many other functions that are currently being evaluated and that any function can have an associated environment. These two additional complexities can sometimes be slightly confusing.

When a function is evaluated, a new environment or *frame* is created specifically for that evaluation. The global environment is always recorded as frame 0 and other frames count up from there. The frame provides bindings between the formal arguments for the function and the user-supplied values. It is also where any local variables have their bindings stored.

The *parent frame* of a function evaluation is the environment from which the function was invoked or called. It is not necessarily numbered one less than the frame number of the current evaluation, although that is usually the case. Symbols in the parent frame have no effect on evaluation of the current function. In some programming languages these symbols do and the language is often said to have dynamic scope, but this is not the case in R.

However, programmers have access to the entire call stack and virtually all objects and frames that are defined on it. Access is obtained through a set of related functions all with the prefix `sys.`; `sys.frame` and `sys.parent` are the the more commonly used. Programmers should avoid the use of these tools as they can make it difficult to understand or reason about a program. One of the common idioms is the use of `sys.frame(sys.parent())` to obtain access to variable bindings that are present in the environment that the current function has been invoked from. For this task, the special syntax of `parent.frame` should be preferred.

2.12.4 Indirect function invocation

Functions can be passed as arguments to functions and evaluated directly, but this is sometimes not convenient or, in some cases, the appropriate function name must be constructed. We now consider the problem of invoking a function when only a character string is available. If all arguments are known, then one can make use of the function `get` as is shown in the code chunk below.

```
> b = get("foo")
> b(23)

[1] 33
```

In other cases it may not be possible to directly invoke the function. In such cases, `do.call` can be used. The arguments to `do.call` are the name of the function, or a function itself and an optionally named `list` of arguments. Argument matching is carried out in a manner analogous to that of the usual argument matching.

Another mechanism for controlling evaluation is the function `with`. The syntax is `with(data, expr)`, where `data` can be a list, an environment or a data frame. In all cases, the values must be named and `expr` is evaluated in a specially created evaluation environment whose parent frame is the evaluation environment that `with` was called from. All variables named in `data` are bound in the environment.

One thing to note is that any assignments made in `expr` take place in the specially constructed evaluation environment and hence are local. They will be discarded at the completion of the call to `with` unless they have been assigned into a more permanent memory location. `with` is most often useful for evaluation of formulae and other modeling functions where the current evaluation semantics are somewhat peculiar.

2.12.5 Evaluation on exit

In many situations it is helpful to be able to ensure that certain values are reset or file handles closed when exiting from a function. Typically these actions should happen whether the function is exited normally or via an error or other condition. The `on.exit` can be used to establish a set of expressions that will be evaluated when the function containing the call to `on.exit` exits.

`on.exit` takes two arguments: the expression to be evaluated and `add`, which must be a logical variable. If `add` is `TRUE`, then the expression is added to the list of expressions to be evaluated; and if `FALSE`, then the provided expression replaces the currently established expression.

2.12.6 Other topics

One should be careful to differentiate between a side effect and a value. A simple example of that difference comes from the `print` function. In the following code we will use `print` to print the value of an object, which is a side effect, and then we will show that the return value of the call to `print` is the object itself. Hence, we have both a side effect and a value. All R functions return values; sometimes the value is `NULL`, but a value is always returned. If the last statement in the body of the function is a call to `invisible`, then the value will not be printed, but it is still returned.

The code below is very simple; first we print the string a and notice that it is printed, but that the function `print` does not print any value after returning control to the console. However, if we assign the return value from `print`, we see that there is one and that it is the value that was printed.

```
> print("a")

[1] "a"

> v = print("a")

[1] "a"

> v

[1] "a"
```

The function `eval` evaluates an expression in an environment. The user can provide an environment or make use of the default values. Since `eval` evaluates its first argument, it can be problematic to evaluate complex expressions and very often these are wrapped in a call to `quote` that, when evaluated, simply returns its argument. The alternate form, `evalq`, does the quoting automatically, and this is the more common form of usage.

Notice the difference in evaluation of the quoted argument and the argument itself. We first assign to the symbol x the expression `1:10`. Then when we call `eval(x)`, the first evaluation is to replace x with its value; the second evaluation is to evaluate that value and we see that the output is the vector of integers from 1 to 10. In the call to `evalq`, there is (at least conceptually) no first evaluation and so the evaluation that does occur is the replacement of the symbol x with its value, and that is what is printed in the call to `evalq`.

```
> x = expression(1:10)
> x
```

```
expression(1:10)

> eval(x)

 [1]  1  2  3  4  5  6  7  8  9 10

> evalq(x)

expression(1:10)

> eval(quote(x))

expression(1:10)
```

Often, you would like the evaluation to take place in the context of a particular environment. This can be handled by supplying the argument as an additional argument to `eval`. Note that we are not evaluating in that environment, but rather, using it as the first location to look for bindings between symbols in the expression provided as the first argument to `eval`.

```
> e = new.env()
> e$x = 10
> evalq(x, envir = e)

[1] 10
```

The function `local` provides a form of encapsulation that is related to the `let` capabilities in Lisp and Lisp-like languages. `local` evaluates an expression in a specially constructed environment and hence all bindings and changes to bindings are kept within that environment.

In the code segment below, essentially taken from the corresponding R manual page, we demonstrate the construction of mutually recursive functions. The value associated with `gg` is the value of the last expression, which is an anonymous version of `f`. This function has as its evaluation environment the specially constructed environment that was created by the call to `local`. In that environment there are two bindings, one for `f` and one for `k`. Both of these are functions and both have this environment as their evaluation environment and hence they are mutually recursive.

```
> gg = local({
+     k = function(y) f(y)
+     f = function(x) if (x)
```

```
+          x * k(x - 1)
+      else 1
+ })
> gg

function (x)
if (x) x * k(x - 1) else 1
<environment: 0x2edff6c>

> ls(environment(gg))

[1] "f" "k"

> for (i in 1:5) print(gg(i))

[1] 1
[1] 2
[1] 6
[1] 24
[1] 120
```

The use of `local` ensures that the correct version of `f` is found when `k` is invoked. This process is often referred to as name space management and we consider that topic in more detail in the next section.

2.12.7 Name spaces

Name spaces play an important role in good software design in R. A name space is typically associated with a package. The use of a name space allows the author to explicitly import symbols and their bindings from other packages as well as to explicitly export symbols and their values. Users of a package with a name space should only use those symbols that are explicitly exported.

When a package with a name space imports bindings from another package, that second package is not, generally, placed on the search path. This can greatly reduce the amount clutter on the search path and can further help alleviate difficulties encountered by masking, or as it is sometimes called *shadowing*. A detailed description of name spaces and how to implement them in R can be found in Tierney (2003) and R Development Core Team (2007c). Here we consider only their impact on the evaluation process.

A name space allows the programmer to explicitly control the bindings between symbols and values. For example, the symbol `pi` is defined in the **base**, but it could be inadvertently overridden by an assignment in the user's workspace, perhaps referring to p sub i, with some potential for unintended results.

A name space changes the evaluation process. When a symbol is being sought in a name space, first the internal definitions are searched, second any explicitly imported symbols are considered and finally the **base** is considered. After that, the usual rules, using the search path, are followed.

A registry of loaded name spaces is maintained and can be examined using the loadedNamespaces function.

Accessing symbols, or variables, exported by a package with a name space can be done using a fully qualified variable reference. Fully qualified variable references consist of the package name and the variable name separated by a double colon. Exported variables may also be accessed by variable name, but would then be subject to masking if other definitions were to precede them on the search path. By making use of the fully qualified variable name, users can ensure that they have obtained the desired binding. When a fully qualified variable name is used, the associated name space is loaded (but not attached). In the code chunk below we query to see which name spaces have been loaded and then access the lda, linear discriminant analysis, in the **MASS** package. Afterwards we again query for the loaded name spaces and then for the current search list and see that while the name space for **MASS** has been loaded, **MASS** is not on the search path.

```
> loadedNamespaces()

 [1] "AnnotationDbi" "Biobase"       "DBI"
 [4] "KernSmooth"    "RColorBrewer"  "RSQLite"
 [7] "annotate"      "base"          "geneplotter"
[10] "grDevices"     "graphics"      "grid"
[13] "lattice"       "methods"       "stats"
[16] "tools"         "utils"         "xtable"

> MASS::lda

function (x, ...)
UseMethod("lda")
<environment: namespace:MASS>

> loadedNamespaces()

 [1] "AnnotationDbi" "Biobase"       "DBI"
 [4] "KernSmooth"    "MASS"          "RColorBrewer"
 [7] "RSQLite"       "annotate"      "base"
[10] "geneplotter"   "grDevices"     "graphics"
[13] "grid"          "lattice"       "methods"
[16] "stats"         "tools"         "utils"
[19] "xtable"

> search()
```

```
 [1] ".GlobalEnv"             "package:geneplotter"
 [3] "package:annotate"       "package:xtable"
 [5] "package:AnnotationDbi"  "package:RSQLite"
 [7] "package:DBI"            "package:lattice"
 [9] "package:Biobase"        "package:tools"
[11] "package:stats"          "package:graphics"
[13] "package:grDevices"      "package:utils"
[15] "package:datasets"       "package:methods"
[17] "Autoloads"              "package:base"
```

While it is possible, and sometimes desirable, to make use of variables from other packages using fully qualified names, it is generally better to import the symbols explicitly in the NAMESPACE file. The main reason for this is that it is possible to determine package dependencies programmatically from the package DESCRIPTION file and NAMESPACE file, but dependencies that arise from fully qualified variable names are much harder to detect. One exception to this rule is when a package wants to make use of functionality from another package only when that other package is available. Then, explicit inclusion of the second package in either the DESCRIPTION file or the NAMESPACE file would cause all users to have the second package available, which may not be the intention.

2.13 Lexical scope

One of the main differences between R and S-Plus is lexical scoping, which R has and S-Plus does not. When properly used, lexical scope provides a powerful mechanism for controlling evaluation and ensuring that the intended sets of bindings between symbols and values are used. We begin with a very simple example that demonstrates the issues.

The code below defines a function named `foo` with no formal arguments. In `foo`, a local variable named `y` is defined and bound to the value 10. And a function is returned. In that function the symbol `y` is used; but since it is not a formal argument to the function, it is a free variable. Then `foo` is evaluated and the return value is stored in the variable named `bar`. Note that `bar` is itself a function. Hence, `bar` can be evaluated. The concern here is: what is an appropriate binding for the symbol `y`? In some computer languages this would be an error, but in many others, the binding for `y` is defined to be that binding that was present (if any) when the function was created. So, in the present example, that binding would be to the value 10 and we see that that is consistent with the output.

```
> foo = function() {
+      y = 10
+      function(x) x + y
+ }
> bar = foo()
> bar

function (x)
x + y
<environment: 0x2d3c580>

> is.function(bar)

[1] TRUE

> bar(3)

[1] 13
```

Functions, such as `foo` in the preceding example, that have an enclosing environment are often referred to as closures. A closure can either be created, as in that example, by the explicit creation of a function in an environment other than the global environment or they can be created explicitly by attaching an environment to a function using `env =` and then populating that environment, as is shown in the next example.

```
> bar2 = function(x) x + z
> e1 = new.env()
> e1$z = 20
> tryCatch(bar2(11), error = function(x) "bar2 failed")

[1] "bar2 failed"

> environment(bar2) = e1
> tryCatch(bar2(11), error = function(x) "bar2 failed")

[1] 31
```

In R, functions can be used anywhere a value is needed. Functions can be passed to other functions as arguments, and functions can be returned as the value of a function. The use of lexical scope is a predominant method for controlling evaluation in Lisp and Scheme.

We now consider two somewhat more realistic examples adapted from Gentleman and Ihaka (2000). One involves the use of likelihood functions and the other the not unrelated concept of function optimization – typically one is interested in obtaining maximum likelihood estimates.

2.13.1   Likelihoods

Suppose we observe a sample of size n that we believed to be from the Exponential density $f(x) = \mu \exp(-x\mu)$, where both μ and x must be positive. In order to estimate μ, one can use the likelihood principle. The log likelihood function for a sample, $(x_1, \ldots, x_n)$, from an Exponential(μ) distribution is $l(\mu) = n \log(\mu) - \mu \sum(x_i)$. The maximum likelihood estimate is the value of μ that maximizes this function.

Likelihood functions are commonly used in both research and teaching. It would be convenient to have some means of creating a likelihood function. This means that we want to have some function, which we will call a *creator*, that we pass data to and get back a likelihood function. We will call this function the *returned function*. This likelihood function would then take as arguments values of the parameter (μ in the case above) and return the likelihood at that point for the data that was supplied to the creator. To do so, the returned function needs to have access to the values of the data that were passed to the creator.

If the programming language has lexical scope, there is no problem because the returned function is created inside the creator and hence has access to all variable definitions that were in effect at the time that it was created.

In the following example, `Rmlfun` is a creator. It sets up several local variables that will be needed by the likelihood function and whose values depend on the data supplied. Then the likelihood function is created and returned. The environment associated with the returned function is the environment that was created by the invocation of `Rmlfun`, which means that the variables `n` and `sumx` will have bindings in that environment.

```
> Rmlfun = function(x) {
+       sumx = sum(x)
+       n = length(x)
+       function(mu) n * log(mu) - mu * sumx
+ }
```

Subsequent evaluation of `Rmlfun` causes the creation of a new environment with bindings to `n` and `sumx` that depend on the arguments supplied to `Rmlfun`. This environment does not interfere in any way with any environment created by previous invocations of `Rmlfun`.

```
> efun = Rmlfun(1:10)
> efun(3)

[1] -154.0139

> efun2 = Rmlfun(20:30)
> efun2(3)

[1] -812.9153

> efun(3)

[1] -154.0139
```

2.13.2   Function optimization

In this section we extend the example given above to indicate one of the areas where lexical scope can provide great simplifications of the code. We will use simple examples and naive implementations of them so that the points regarding lexical scope are not lost amid the complexity of function optimization. For the reader this can be paraphrased as: do not use these methods; they are only examples and there are better ways to solve these problems. However, even the better solutions benefit from lexical scope so we lose nothing and gain simplicity for our purpose.

Optimization problems frequently arise in all areas of statistics and one common problem is in finding the maximum likelihood estimate. In many cases the likelihood is convex in the parameters and hence has a single maximum. In that case the maximum likelihood estimate can be obtained by finding the place where the score function (the first derivative of the likelihood) is zero.

One simple method for finding the zero of an arbitrary function, $f(x)$, of one variable is Newton's method. If a is an initial guess as to the value of x such that $f(x) = 0$, then an improved guess is obtained via

$$x_{new} = a - f(a)/f'(a). \tag{2.1}$$

This process can then be iterated until a value of x_{new} is obtained such that $f(x_{new})$ is sufficiently close to zero.

In most of the problems that arise in statistics, the function being optimized depends not only on the parameter that we are optimizing over, but also on many other variables (usually the data). Because of that, one can never really use the simple form of Equation 2.1. In most implementations there must be some means of passing the extra information to the optimizer. This generally complicates the code and often results in a solution that is not easily extended.

However, when the language has lexical scope, the simple form can be used for many problems. Consider the slightly extended likelihood function generator given below.

```
> Rmklike = function(data) {
+     n = length(data)
+     sumx = sum(data)
+     lfun = function(mu) n * log(mu) - mu *
+         sumx
+     score = function(mu) n/mu - sumx
+     d2 = function(mu) -n/mu^2
+     list(lfun = lfun, score = score, d2 = d2)
+ }
```

In this function we return not only the likelihood function, but also functions to obtain the score and the second derivative.

The optimizer can then be written in the following way:

```
> newton = function(lfun, est, tol = 1e-07, niter = 500) {
+     cscore = lfun$score(est)
+     if (abs(cscore) < tol)
+         return(est)
+     for (i in 1:niter) {
+         new = est - cscore/lfun$d2(est)
+         cscore = lfun$score(new)
+         if (abs(cscore) < tol)
+             return(new)
+         est = new
+     }
+     stop("exceeded allowed number of iterations")
+ }
```

The function `newton` can be used to find the zero of any univariate function provided that the function passed in adheres to the protocol that the zero function is stored in the list as `score` and its derivative is stored in the list as d2.

2.13.2.1 Other considerations

Lexical scope is implemented by associating an environment with a function. That environment can contain bindings for any unbound symbols in the body of the function, and these bindings will be used first when the function is evaluated.

R Programming for Bioinformatics

Environments can be directly assigned to functions and values can be inserted directly into those functions for use. Standard tools for assigning and changing values in environments can be used, and work as documented.

In the code chunk below we first create an environment e1 and bind the symbol a to the value 10. Next the function foo is defined; it has one formal argument, x, and one unbound value, a. Then e1 is assigned as the environment of foo. Now, when foo is evaluated, e1 will be used to provide a binding for the symbol a.

```
> e1 = new.env()
> e1$a = 10
> foo = function(x) x + a
> environment(foo) = e1
> foo(4)

[1] 14
```

The environment e1 can be manipulated directly, as we see in the next code chunk. The value associated with the symbol a can be changed and that change is propagated.

```
> e1$a = 20
> foo(4)

[1] 24

> e1[["a"]]

[1] 20
```

2.14   Graphics

One of the real strengths of the R language is its comprehensive graphics capabilities. Since these have substantial supporting documentation, and there are books devoted to the graphics systems, (Murrell, 2005; Sarkar, 2008), we give them only a cursory treatment here, and recommend that the interested reader consult the resources cited here for a more substantial treatment of the topics.

There are three different systems that can be used, with some substantial overlap between them. There is a basic, or old-style graphics system, and a newer system called **grid** that gives more control over some tools. These two graphics systems are well documented in Murrell (2005). The function demo can be used to see some online examples and the **grid** package has a number of vignettes.

The third system of note is implemented in the **lattice** package. It primarily provides tools to help visualize a number of related plots simultaneously. There is a book (Sarkar, 2008) and a demo that can be accessed via the demo function.

The R graphics system is device oriented. At any one time, a single device is active and accepts input. Some devices (typically those that render on a computer screen) also have some interactive capabilities. In particular, the functions *locator* and *identify* are available. During an interactive session, it is possible to have several on-screen devices presenting different plots, but at any one time, only one is active.

There are other devices, such as pdf, for producing documents in the portable document format, postscript for producing plots in Postscript, as well as bitmap, xfig and pictex devices are always available. Other devices such as X11, png and jpg will be available if the necessary software has been installed on the corresponding machine. To initialize a device, one simply invokes the appropriate function, and from that point on all plotting commands are directed to the new device, at least until it is closed or another device opened.

In order to navigate the different devices, there are a number of different functions that can be used; these are listed below.

dev.cur returns the identity of the active device.

dev.list lists all active devices.

dev.next makes the next device in the list active.

dev.set makes the specified device active.

dev.copy copies the graphics contents of the active device to the specified device.

In order to view multiple plots simultaneously, users can either maintain multiple active devices, or place multiple plots on the same device. The graphics parameters (see the manual page for par for more details) mfrow and mfcol can be used to set up the active graphics device so that multiple plots will appear. For more complex arrangements, see the heatmap function for one example, use the layout command to obtain more control over the shapes and size of the plotting regions.

The graphics parameters control many different aspects of how plots are rendered, including setting margins, controlling whether a plot is overlayed on an existing plot, and whether the user should be queried for input before erasing a plot. There are far more parameters than we can easily describe, and interested readers are encouraged to explore these different settings themselves.

Exercise 2.21

- *Produce a bitmap image of a plot. Which parameters must you set? Which parameters are optional?*

- *Use `layout` to create a scatterplot with histograms on the sides. Hint: see the manual page.*

- *Use `dev.copy` to copy this to a PDF device and then open the resulting PDF document using your favorite viewer.*

- *What does the graphics parameter `cex` do?*

- *Can you find the size of the figure? What units (e.g., pixels, inches, etc.) can this be obtained in?*

Chapter 3

Object-Oriented Programming in R

3.1 Introduction

Object-oriented programming (OOP) has become a widely used and valuable tool for software engineering. Much of its value derives from the fact that it is often easier to design, write and maintain software when there is some clear separation of the data representation from the operations that are to be performed on it. In an OOP system, real physical things (like airline passengers or the data from a microarray experiment) are generally represented by classes, and methods (functions) are written to handle the different manipulations that need to be performed on the objects.

The views that many people have of OOP have been based largely on exposure to languages like Java, where the system can be described as class-centric. In a class-centric system, classes define objects and are repositories for the methods that act on those objects. In contrast, languages such as Dylan (Shalit, 1996), Common Lisp (Steele, 1990), and R separate the class specification from the specification of generic functions, and could be described as being function-centric systems.

R currently supports two internal OOP systems, and several others are available as add-on packages. In this chapter we discuss the two internal systems. The first, called S3, is documented in Chambers and Hastie (1992) while the second, S4, was first described in Chambers (1998) and later updated in Chambers (2008). R has become very popular and is now being used for projects that require substantial software engineering as well as its continued widespread use as an interactive environment for data analysis. This essentially means that there are two masters – *reliability* and *ease of use*. S3 is indeed easy to use, but can be made unreliable through nothing other than bad luck, or a poor choice of names, and hence is not a suitable paradigm for constructing large systems. S4, on the other hand, is better suited for developing large software projects but has an increased complexity of use.

Freidman et al. (2001) list four general elements that an object-oriented programming language should support.

objects: encapsulate state information and control behavior.

classes: describe general properties for groups of objects.

inheritance: new classes can be defined in terms of existing classes.

polymorphism: a (generic) function has different behaviors, although similar outputs, depending on the class of one or more of its arguments.

Virtually every OOP language implements these in different ways. In S3, there is no formal specification for classes and hence there is, at best, weak control of objects and inheritance. The emphasis of the S3 system was on generic functions and polymorphism. In S4, formal class definitions were included in the language and based on these, more controlled software tools and paradigms for the creation of objects and the handling of inheritance were introduced.

We also note that when using OOP, much of the important detail in the programs are contained in the class hierarchy and in understanding which classes have specific methods available for them. Thus, tools for inspecting and visualizing this structure are invaluable in understanding how a program functions. We touch on this topic in Section 3.9.

3.2   The basics of OOP

One can separate a discussion of OOP into two related but distinct sets of concepts. First are the classes, which describe the objects that will be represented in computer code. A class specification details all the properties that are needed to describe an object. An object is an instance of exactly one class and it is the class definition and representation that determine the properties of the object. Instances of a class differ only in their state. New classes can be defined in terms of existing classes through an operation called inheritance. Inheritance allows new classes to extend, often by adding new slots or by combining two or more existing classes into a single composite entity. If a class A extends the class B, then we say that A is a superclass of B, and equivalently that B is a subclass of A. No class can be its own subclass. A class is a subclass of each of its superclasses.

If the language only allows a class to extend, at most, one class, then we say that language has single inheritance. Computing the class hierarchy is then very simple, since the resulting hierarchy is a tree and there is a single unique path from any node to the root of the tree. This path yields the class linearization. In the S3 system, the class of an instance is determined by the values in the `class` attribute, which is a vector, and hence is also linear. If the language allows a class to directly extend several classes, then we say that the language supports multiple inheritance and computing the class linearization is more difficult. S4 supports multiple dispatch.

A method is a type of function that is invoked depending on the class of one or more of its arguments and this process is called dispatch. While in some systems, such as S3, methods can be invoked directly, it is more common for them to be invoked via a generic function. When a generic function is invoked, the set of methods that might apply must be sorted into a linear order, with the most specific method first and the least specific method last. This is often called method linearization and computing it depends on being able to linearize the class hierarchy. If the language supports dispatching on a single argument, then we say it has single dispatch. Both Java and the S3 system use single dispatch. When the language supports dispatching on several arguments, we say that the language supports multiple dispatch and the set of specific classes of the arguments for each formal parameter of the generic function is called the signature. S4 supports multiple dispatch. With multiple dispatch, the additional complication of precedence of the arguments arises. In particular, when method selection depends on inheritance, there may be more than one superclass for which a method has been defined. In this case, a concept of the distance between the class and its superclasses is used to guide selection; more details on method linearization are given in Section 3.2.2.

The evaluation process for a call to a generic function is roughly as follows. The actual classes of supplied arguments that match the signature of the generic function are determined. Based on these, the available methods are ordered from most specific to least. Then, after evaluating any code supplied in the generic, control is transferred to the most specific method. In S4, a generic function has a fixed set of named formal arguments and these form the basis of the signature. Any call to the generic will be dispatched with respect to its signature. There can be arguments to the generic that are not part of the signature and are not used to determine dispatch. Provided the generic function uses the ... argument, methods can have fairly arbitrary non-signature arguments.

Single inheritance and single dispatch yields an easy to understand and easy to implement paradigm that solves about 90% of all programming problems and hence is popular; but when it is not sufficient, the convolutions needed to overcome its deficiencies can be substantial. For example, the visitor pattern described in Gamma et al. (1995) is essentially a mechanism to support multiple dispatch in a language with only single inheritance.

One of the advantages of an OOP paradigm is that relatively little checking of the input values is needed. The reason is that the class of an object is known if we dispatch on it. And since instances of a class differ only in their state (i.e., they generally have the same slots and the same classes of values in those slots), we can write the method with very strong assumptions about the inputs.

3.2.1 Inheritance

One of the advantages of a class system is the concept of inheritance. Consider as an example the modeling of airline passengers. One implementation would define the *Passenger* class as having slots for passenger name (which might itself be a class), an origin and a destination. Now, to implement a new class for frequent flyers, e.g., *FreqFlyer*, we do not want to create a whole new set of class definitions, but rather we can extend the *Passenger* by adding one or more slots to describe new properties that will be recorded only for frequent flyers. We then say that the *FreqFlyer* is a subclass of *Passenger* and that *Passenger* is a superclass of *FreqFlyer*.

The inheritance relationships imply a form of polymorphism. Any instance of the subclass, in our example the *FreqFlyer*, can be used in place of an instance of the superclass, in our example *Passenger*. This must be true since the *FreqFlyer* class has every slot that an instance of the *Passenger* class has. The relationship between a subclass and its superclasses should be an *is a* relationship. Every frequent flyer *is a* passenger and not all passengers are frequent flyers.

Sometimes the notion of subclass and superclass can be confusing. One reason that the more specialized class is called a subclass is because the set of objects that can be used exchangeably with the *FreqFlyer* class are a subset of those that can be used exchangeably with the *Passenger* class. In the example below, we provide a very basic S4 implementation of the *Passenger* and *FreqFlyer* classes.

```
> setClass("Passenger", representation(name = "character",
+      origin = "character", destination = "character"))

[1] "Passenger"

> setClass("FreqFlyer", representation(ffnumber = "numeric"),
+      contains = "Passenger")

[1] "FreqFlyer"

> getClass("FreqFlyer")

Slots:

Name:       ffnumber          name        origin
Class:       numeric     character     character

Name:  destination
Class:    character

Extends: "Passenger"

> subClassNames("Passenger")

[1] "FreqFlyer"
```

```
> superClassNames("FreqFlyer")

[1] "Passenger"
```

Exercise 3.1

Define a class for passenger names that has slots for the first name, middle initial and last name. Change the definition of the Passenger class to reflect your new class. Does this change the inheritance properties of the Passenger class or the FreqFlyer class?

3.2.2 Dispatch

We must also briefly digress to consider some of the issues involved in applying methods to objects. A method is a specialized function that can be applied to instances of one or more classes. The process of determining the appropriate method to invoke is called dispatch. A call to a function, such as `plot`, will invoke a method that is determined by the class of the first argument in the call to `plot`.

When a generic function is called, it must examine the supplied arguments and determine the applicable methods. All applicable methods are ordered; details on how this is done for S4 are given in Section 3.4.9, while for S3 the hierarchy is intrinsically linear and hence has an obvious order. In both systems, the applicable methods are arranged from most specific to least specific and the most specific method is invoked. During evaluation, control may be passed to less specific methods by calling `NextMethod` in S3 and via `callNextMethod` for S4. This strategy tends to help simplify the code. If we return to our frequent flyer example, we can imagine a print method for passengers that prints their names and flight details. A print method for frequent flyers could simply invoke the passenger method, and then add a line indicating the frequent flyer number. Using this approach, very little additional code is needed; and if the printing of passenger information is changed, the update is automatically applied to printing of frequent flyer information.

Exercise 3.2

Write a simple show method for the Passenger class. Write a show method for theFreqFlyer class that makes use of the show method for passengers. For S4 you will want to use setMethod and callNextMethod, while for an S3 implementation you will need to use NextMethod and name the print methods print.Passenger and print.FreqFlyer.

With both S3 and S4, dispatching is implemented through the use of generic functions. In the S3 system, the generic function typically only examines the first argument and dispatches depending on its class. In S3, methods are not explicitly registered but are determined by a function naming convention that is described later. S4, on the other hand, requires explicit method registration. In S4, when the generic function is invoked, it examines the classes of all arguments in its signature and then linearizes the methods, invoking the most specific one.

3.2.3   Abstract data types

In some discussions there is confusion between the use of abstract data types (ADT) and OOP. It is useful to realize that the ADT paradigm can be adopted in any language, regardless of whether or not it supports OOP.

Every time a decision is made about how to represent a set of quantities, either simple ones, such as the time of day, or more complex ones, such as the output of a DNA microarray experiment, a new data type is created. The data must be stored in some format and any processing of the data relies on manipulating that data. At some time in the future it may become important to change the format the data are stored in. In order not to have to rewrite all code that manipulates the data, the notion of an ADT can be used. By this we simply mean that we conceptually separated the *representation* of the object from the *interface* to the object. The representation provides specific details for storage of the data, and details of the implementation should not be relied on to access the data. All users of the data type must restrict their operations to those defined by the interface.

It is quite obvious why ADT and OOP are often mistaken for each other. At one level, OOP is merely a set of software tools that help to adopt an ADT approach. A class specification can be thought of as the data representation, while the methods define the interface. But, it is possible to use ADT without a class system.

Consider the following simple example, in S4. Suppose that we have a *Rectangle* class and that this class should respond to requests that ask for the area of the rectangle. We first define a class for rectangle that includes a specific slot for the area. We next define a generic function for area, and define a method for the *Rectangle* class.

```
> setClass("Rectangle", representation(h = "numeric",
+     w = "numeric", area = "numeric"))

[1] "Rectangle"

> myr = new("Rectangle", h = 10, w = 20, area = 200)
> setGeneric("area", function(shape) standardGeneric("area"))

[1] "area"

> setMethod("area", signature(shape = "Rectangle"),
+     function(shape) shape@area)
```

```
[1] "area"

> myr@area

[1] 200

> area(myr)

[1] 200
```

Any user can either access the area directly by accessing the slot with `myr@area` or by calling the `area` generic function. Accessing the slot breaks the data type abstraction; you are relying on the implementation. Using the generic function makes use of the ADT. If the representation were to change to that shown below, any code relying on the generic function will continue to work and any code relying on slot access will fail. By using ADTs, it is simpler to change the representation of data types as a project evolves.

```
> setClass("Rectangle", representation(h = "numeric",
+      w = "numeric"))

[1] "Rectangle"

> setMethod("area", "Rectangle", function(shape) shape@h *
+      shape@w)

[1] "area"
```

3.2.4 Self-describing data

One of the major uses of OOP within the Bioconductor Project is in the construction of self-describing data classes. The most widely used is the *ExpressionSet* class defined in the **Biobase** package. Our goal is to define a self-describing data object that can be used to carry out a reasonable analysis of the data. If all information is stored in a single object it is easier to save it, to share it with others, or to use it as input to a function. This is consistent with the notion that you would like to place all data relevant to the experiment into a single file folder and to place it into a filing cabinet so that later you can find all the information you need in one place.

The data might be stored in either a matrix or a `data.frame`. And while informative row and column labels can be used, it is difficult to encode all relevant information about the variables in the labels. One solution is to create a compound object that holds both the data and the metadata about the variables, and possibly about the samples. Defining a suitable class yields self-describing data.

The major benefits that we have found to programming with self-describing data are that it is easy to return to a project after some months and re-do an analysis. We have also found that it is relatively easy to hand off a project from one analyst to another. But perhaps the greatest benefit has come from defining specialized subsetting methods, that is, methods for [that help to construct an appropriate subset of the object, with all variables correctly aligned.

3.3   S3 OOP

The S3 system is relatively easy to describe and to use. It is particularly well suited to interactive use but is not particularly robust. Generic functions and methods are quite widely used but there is little use of inheritance and classes are quite loosely defined. In some sense, all objects in R are instances of some class. Some classes are *internal* or *implicit* and others are specified explicitly, typically by using the `class` attribute. In the S3 system, one determines the class of an object using the function `class`, and for most purposes this is sufficient; however, there are some important exceptions that arise with respect to internal functions. While there is no formal mechanism for organizing or representing instances of a class, they are typically lists, where the different slots are represented as named elements in the list. Using `setOldClass` will register an S3 class as an S4 class.

The `class` attribute is a vector of character values, each of which specifies a particular class. The most specific class comes first, followed by any less specific classes. For our frequent flyer example from Section 3.2.1, the class vector should always have `FreqFlyer` first and `Passenger` second. The recommended way of testing whether an S3 object is an instance of a particular class is to use the `inherits` function. Direct inspection of the class attribute is not recommended since implicit classes, such as *matrix* and *array*, are not listed in the class attribute. Notice in the code below that the class of x changes when a dimension attribute is added, that there is no `class` attribute, and that once x is a *matrix* it is no longer considered to be an *integer*.

```
> x = 1:10
> class(x)

[1] "integer"

> dim(x) = c(2, 5)
> class(x)
```

```
[1] "matrix"

> attr(x, "class")

NULL

> inherits(x, "integer")

[1] FALSE
```

In the next example we return to our *FreqFlyer* example and provide an S3 implementation.

```
> x = list(name = "Josephine Biologist", origin = "SEA",
+     destination = "YXY")
> class(x) = "Passenger"
> y = list(name = "Josephine Physicist", origin = "SEA",
+     destination = "YVR", ffnumber = 10)
> class(y) = c("FreqFlyer", "Passenger")
> inherits(x, "Passenger")

[1] TRUE

> inherits(x, "FreqFlyer")

[1] FALSE

> inherits(y, "Passenger")

[1] TRUE
```

A major problem with this approach is that there is no mechanism that programmers can use to ensure that all instances of the *Passenger* or *FreqFlyer* classes have the correct slots, the correct types of values in those slots, and the correct class attribute. One can easily produce an object with these classes that has none of the *slots* we have defined. And as a result, one typically has to do a great deal of checking of arguments in every S3 method.

The function `is.object` tests whether or not an R object has a `class` attribute. This is somewhat important as the help page for `class` indicates that some dispatch is restricted to objects for which `is.object` is true.

```
> x = 1:10
> is.object(x)
```

```
[1] FALSE

> class(x) = "myint"
> is.object(x)

[1] TRUE
```

3.3.1   Implicit classes

The earliest versions of the S language predate the widespread use of object-oriented programming and hence the class representations for some of the more primitive or basic classes do not use the `class` attribute. For example, functions and closures are implicitly of class *function* while matrices and arrays are implicitly of classes *matrix* and *array*, respectively.

```
> x = matrix(1:10, nc = 2)
> class(x) = "matrix"
> x

     [,1] [,2]
[1,]    1    6
[2,]    2    7
[3,]    3    8
[4,]    4    9
[5,]    5   10

> is.object(x)

[1] FALSE

> oldClass(x) = "matrix"
> x

     [,1] [,2]
[1,]    1    6
[2,]    2    7
[3,]    3    8
[4,]    4    9
[5,]    5   10
attr(,"class")
[1] "matrix"

> is.object(x)

[1] TRUE
```

Exercise 3.3

The S3 system has been used for some years and a very extensive set of tools for statistical modeling has been developed based on this system (Chambers and Hastie, 1992). Among the builtin classes is glm. *Fit a simple generalized linear model (using an example from the help page for* glm *is the easiest way) and examine its structure. What classes does* glm *extend? What are the slots in a* glm *instance?*

3.3.2 Expression data example

In this section we consider constructing something similar to the *ExpressionSet* class used in Bioconductor entirely within the S3 system. We first define a class that will relate variable names and their descriptions. This will be a named vector, where the names correspond to the names we will use for the variables in the `data.frame` and the values are the textual descriptions of the variables. In the code below we create such an object and call it class *VARLS3*.

```
> ex1VL = c("Sex, M=MALE, F=FEMALE", "Age in years")
> names(ex1VL) = c("Sex", "Age")
> class(ex1VL) = "VARLS3"
```

Next we simulate data for $n = 10$ samples and $G = 100$ genes. We first set the seed to ensure reproducibility.

```
> set.seed(123)
> simExprs = matrix(rgamma(10000, 500), nc = 10,
+      nr = 100)
> simS = sample(c("M", "F"), 10, rep = TRUE)
> simA = sample(30:45, 10, rep = TRUE)
> simPD = data.frame(Sex = simS, Age = simA)
```

Now that we have simulated data, we can construct an instance of our class. For S4 classes there is a builtin function `new` that can be used, but for S3 there is no such mechanism; however, we will write a constructor function as it can then be used to make other instances. One thing to notice is just how much of `new.EXPRS3` is merely checking the types of the inputs; in an S4 implementation, this extensive checking would not be needed for any argument that was dispatched on.

```
> new.EXPRS3 = function(Class, eData, pData,
+     cDesc) {
+     if (!is.matrix(eData))
+         stop("invalid expression data")
+     if (!is.data.frame(pData))
+         stop("invalid phenotypic data")
+     if (!inherits(cDesc, "VARLS3"))
+         stop("invalid cov description")
+     ncE = ncol(eData)
+     nrP = nrow(pData)
+     if (ncE != nrP)
+         stop("incorrect dimensions")
+     pD = list(pData = pData, varLabels = cDesc)
+     class(pD) = "PHENODS3"
+     ans = list(exprs = eData, phenoData = pD)
+     class(ans) = class(Class)
+     ans
+ }
```

And we can create new instances of the class *EXPRS3* by calling the function `new.EXPRS3`.

```
> myES3 = new.EXPRS3("EXPRS3", simExprs, simPD,
+     ex1VL)
```

Readers should treat this example as being solely pedagogical; the *ExpressionSet* class in the **Biobase** package provides a much richer and more extensive implementation of these ideas.

3.3.3   S3 generic functions and methods

In S3, the generic function is responsible for setting up the evaluation environment and for initiating dispatch. A generic function does this through a call to `UseMethod` that initiates the dispatch on a single argument, usually the first argument to the generic function. The generic is typically a very simple function with only two formal arguments, one often named x and the other the ... argument. If the ... argument is not used in the generic, then no method can have a formal argument that is not also a formal argument of the generic. Thus, it is good practice for all methods to include all arguments to the generic and for all generic functions to include the ... argument since this ensures that methods that may be added later have sufficient flexibility to add new arguments that are appropriate to the computations they will perform. A disadvantage of this approach is that mistakes in naming arguments will be silently ignored. The mis-typed name will not match any formal argument and hence is placed in the ... argument, where it is never used.

In R, UseMethod dispatches on the class as returned by class, not that returned by oldClass. Not all method dispatch honors implicit classes. In particular, group generics (Section 3.3.5) and internal generics do not. Group generics dispatch on the oldClass for efficiency reasons, and internal generics only dispatch on objects for which is.object is TRUE. An internal generic is a function that calls directly to C code (a primitive or internal function), and there checks to see if it should dispatch. To make use of these, you will need to explicitly set the class attribute. You can do that using class<-, oldClass<- or by setting the attribute directly using attr<-.

For most generic functions, a default method will be needed. The default method is invoked if no applicable methods are found, or if the least specific method makes a call to NextMethod.

Methods are regular functions and are identified by their name, which is a concatenation of the name of the generic and the name of the class that they are intended to apply to, separated by a dot. A simple generic function named fun and a default method are shown below. The string default is used as if it were a class and indicates that the method is a default method for the generic.

```
> fun = function(x, ...) UseMethod("fun")
> fun.default = function(x, ...) print("In the default method")
> fun(2)

[1] "In the default method"
```

Next, consider a class system with two classes, *Foo* which extends *Bar*. Then we define two methods: fun.Foo and fun.Bar. We have them print out a message, call the function NextMethod and then print out a second message.

```
> fun.Foo = function(x) {
+     print("start of fun.Foo")
+     NextMethod()
+     print("end of fun.Foo")
+ }
> fun.Bar = function(x) {
+     print("start of fun.Bar")
+     NextMethod()
+     print("end of fun.Bar")
+ }
```

Now we can show how dispatch occurs by creating an instance that has both classes and calling `fun` with that instance as the first argument.

```
> x = 1
> class(x) = c("Foo", "Bar")
> fun(x)

[1] "start of fun.Foo"
[1] "start of fun.Bar"
[1] "In the default method"
[1] "end of fun.Bar"
[1] "end of fun.Foo"
```

Notice that the call to `NextMethod` transfers control to the next most specific method. This is one of the benefits of using an OOP paradigm. Typically, less code needs to be written, and it is easier to maintain as the methods for *Foo* do not need to know much about *Bar* and vice versa, as a specific method for that class can handle the computations.

Exercise 3.4
Returning to our ExpressionSet example, Section 3.3.2, instances of EXPRS3 can be very large and we want to control the default information that is printed by R. Write S3 `print` methods for the PHENODS3 and EXPRS3 classes.

3.3.3.1   Finding methods

Due to the somewhat simple nature of the S3 system, there is very little introspection or reflection possible. The function `methods` reports on all available methods for a given generic function but it does this simply by looking at the names. We demonstrate its use on the S3 generic function `mean` in the code below.

```
> methods("mean")

[1] mean.Date       mean.POSIXct    mean.POSIXlt
[4] mean.data.frame mean.default    mean.difftime
```

One can also use `methods` to find all available methods for a given class. In the code below we find all methods for the class *glm*.

```
> methods(class = "glm")
```

```
 [1] add1.glm*            anova.glm
 [3] confint.glm*         cooks.distance.glm*
 [5] deviance.glm*        drop1.glm*
 [7] effects.glm*         extractAIC.glm*
 [9] family.glm*          formula.glm*
[11] influence.glm*       logLik.glm*
[13] model.frame.glm      predict.glm
[15] print.glm            residuals.glm
[17] rstandard.glm        rstudent.glm
[19] summary.glm          vcov.glm*
[21] weights.glm*

Non-visible functions are asterisked
```

To retrieve the definition of a method, even those that are not exported from a name space, the function `getS3method` can be used, as can the more general function `getAnywhere`.

There is no simple way to determine which S3 classes are defined nor much about those classes.

3.3.4 Details of dispatch

This section provides a detailed discussion of how S3 dispatch works, and can be skipped by readers who are not interested in the inner workings of that system.

As we noted above, methods are identified based solely on their names so a function named `plot.Foo` would be interpreted as a plot method for objects from the class *Foo*, whether or not that is what the author of that function intended. This can lead to problems, as different package authors may use what they believe are perfectly innocent function names, such as `plot.Foo`, never intending for them to be dispatched on. For this reason it is advised that you not use function names with an embedded '.' unless they are intended to be S3 methods.

S3 dispatch works essentially as follows. A call to the function `UseMethod` finds the most specific method and creates a new function call with arguments in the same order and with the same names as they were supplied to the generic function. Any local variables defined in the body of the generic function, before the call to `UseMethod`, are retained in the evaluation environment. Any statements in the body of the generic function after the call to `UseMethod` will not be evaluated as `UseMethod` does not return. `UseMethod` dispatches on the value returned by `class`.

In the example below we redefine the `Foo` method for the function `fun` in order to demonstrate that some special variables have been installed into the evaluation environment of the method.

```
> fun.Foo = function(x, ...) print(ls(all = TRUE))
> y = 1
> class(y) = c("Foo", "Zip", "Zoom")
> fun(y)

[1] "..."                ".Class"
[3] ".Generic"           ".GenericCallEnv"
[5] ".GenericDefEnv"     ".Method"
[7] "x"
```

We examine three of these special variables in a bit more detail in the code
example below. They are .Class, .Generic and .Method; all special variables
are documented and you may make use of them in your code if you wish or
need to – but be careful.

```
> fun.Foo = function(x, ...) {
+     print(paste(".Generic =", .Generic))
+     print(paste(".Class =", paste(.Class,
+         collapse = ", ")))
+     print(paste(".Method =", .Method))
+ }
> fun(y)

[1] ".Generic = fun"
[1] ".Class = Foo, Zip, Zoom"
[1] ".Method = fun.Foo"
```

NextMethod invokes the next most specific method as determined by the
class attribute of the first argument to the generic function. This is achieved
by creating a special call frame for that method. The arguments will be the
same in number, order and name as those to the current method but their
values will be promises to evaluate their name in the current method and
environment. Any arguments matched to ... are handled specially. They are
passed on as the promise that was supplied as an argument to the current
environment. If they have been evaluated in the current (or a previous environment), they remain evaluated. Since NextMethod relies on some of the
special variables described above to determine dispatch, any function that
contains a call to NextMethod should not be invoked directly.

Name spaces affect the availability of methods and generic functions, and
are described more fully in Section 7.3.4. But briefly, S3 methods are exported
from a name space by using the S3Method directive in the NAMESPACE file.

Group	Functions
Math	`abs, acos, acosh, asin, asinh, atan,`
	`atanh, ceiling, cos, cosh, cumsum, exp,`
	`floor, gamma, lgamma, log, log10, round,`
	`signif, sin, sinh, tan, tanh, trunc`
Summary	`all, any, max, min, prod, range, sum`
Ops	`+, -, *, /, ^, < , >, <=, >=, !=, ==, %%,`
	`%/%, &, |, !`

Table 3.1: Group generic functions.

Generic functions require no special markup, but must be exported if they are intended for others to use.

3.3.5 Group generics

The S3 object system also has the capability for defining methods for groups of functions simultaneously. These tools are mainly used to define methods for three defined sets of operators. For several types of builtin functions, R provides a dispatching mechanism for operators. This means that operators such as `==` or `<` can have their behavior modified for members of special classes. The functions and operators have been grouped into three categories and group methods can be written for each of these categories. There is currently no mechanism to add groups. It is possible to write methods specific to any function within a group and then a method defined for a single member of group takes precedence over the group method.

The three groups of operators (Table 3.1) are called `Math`, `Summary` and `Ops`. The online help system provides more detail, as do R Development Core Team (2007b), Chambers and Hastie (1992), Venables and Ripley (2000), and Chambers (2008).

Determining which method to use for operators in the `Ops` group is determined as follows. If both operands correspond to the same method or if one operand corresponds to a method that takes precedence over that of the other operand, then that method is used. If both operands have methods and the methods are conflicting, then the default method is used. If either operand has no corresponding method, then the method for the other operand is used. Class methods dominate group methods.

3.3.6 S3 replacement methods

It is possible in R to have a complex statement as the left-hand side of an assignment, and such an assignment is referred to as a replacement function; see Section 2.7 for more details. The general idea is easily extended to generic functions and methods, and there are very many replacement methods already available in R. In the code below we want to find all replacement functions that have been written for the $ operator. The generic function is $<-, and it is an internal generic. In order to write a replacement function, you must determine the names of the arguments to the generic. This is somewhat tricky, and perhaps the easiest way is to find another assignment method and copy it. The last argument to the assignment version is always named `value`.

In the example below we request all methods for the $<- replacement function and find $<-.`data.frame`, which is a replacement method for objects of class *data.frame*.

```
> methods("$<-")

[1] $<-.data.frame
```

Exercise 3.5
Write a replacement method for the following problem. Let x be a matrix with named rows. Define x$a = y to mean that the row of x named a be set to y. Because $ is an internal generic, it will only dispatch on objects for which is.object is TRUE, so you will need to set the oldClass.

3.4   S4 OOP

The S4 system was designed to overcome some of the deficiencies of the S3 system as well as to provide other functionality that was simply missing from the S3 system. Some of the tensions that arise from mixing the two are discussed in Section 3.8. Among the major changes between S3 and S4 are the explicit representation of classes, together with tools that support programmatic inspection of the class definitions and properties. Multiple dispatch is supported in S4, but not in S3, and S4 methods are registered directly with the appropriate generic. These changes greatly increase the stability of the system and make it much more likely that code will perform as intended by its authors. This comes with some costs, however; code is slightly slower (since all aspects are slightly more complex) and it is more difficult to design and modify a system interactively.

The discussion is separated into two main parts. In the first, the implementation and tools for S4 classes are discussed and in the second, generic functions and methods are considered. Subsequent to that are brief discussions on using S4 methods in packages (more details are presented in Chapter 7) and documentation.

3.4.1 Classes

A class definition specifies the structure, inheritance and initialization of instances of that class. A class is defined by a call to the function `setClass`. Classes are instances of the *classRepresentation* class, and are first-class objects in the language. They can be created by users and existing classes can typically be extended or subclassed. Classes can be *instantiable* or *virtual*; instances can be created for instantiable classes but not for virtual classes. The following arguments can be specified (there are others as well) in the call to `setClass`:

Class a character string naming the class.

representation a named vector of types or classes. The names correspond to the slot names in the class and the types indicate what type of value can be stored in the slot.

contains a character vector of class names, indicating the classes extended or subclassed by the new class.

prototype an object (usually a list) providing the default data for the slots specified in the representation.

validity a function that checks the validity of instances of the class. It must return either `TRUE` or a character vector describing how the object is invalid.

Once a class has been defined by a call to `setClass`, it is possible to create instances of the class through calls to `new`. The `prototype` argument can be used to define default values to use for the different components of the class. Prototype values can be overridden by expressly setting the value for the slot in the call to `new`.

In the code below, we create a new class named *A* that has a single slot, `s1`, that contains numeric data and we set the prototype for that slot to be 0.

```
> setClass("A", representation(s1 = "numeric"),
+     prototype = prototype(s1 = 0))

[1] "A"

> myA = new("A")
> myA

An object of class "A"
Slot "s1":
[1] 0
```

```
> m2 = new("A", s1 = 10)
> m2

An object of class "A"
Slot "s1":
[1] 10
```

We can create a second class B that *contains* A, so that B is a direct subclass of A or, put another way, B inherits from class A. Any instance of the class B will have all the slots in the A class and any additional ones defined specifically for B. Duplicate slot names are not allowed, so the slot names for B must be distinct from those for A.

```
> setClass("B", contains = "A", representation(s2 = "character"),
+     prototype = list(s2 = "hi"))

[1] "B"

> myB = new("B")
> myB

An object of class "B"
Slot "s2":
[1] "hi"

Slot "s1":
[1] 0
```

Classes can be removed using the function `removeClass`. However, this is not especially useful since you cannot remove classes from attached packages. The `removeClass` is most useful when experimenting with class creation interactively. But in most cases, users are developing classes within packages, and the simple expedient of removing the class definition and rebuilding the package is generally used instead. We demonstrate the use of this function on a user-defined class in the code below.

```
> setClass("Ohno", representation(y = "numeric"))

[1] "Ohno"

> getClass("Ohno")
```

```
Slots:

Name:        y
Class: numeric

> removeClass("Ohno")

[1] TRUE

> tryCatch(getClass("Ohno"), error = function(x) "Ohno is gone")

[1] "Ohno is gone"
```

3.4.1.1 Introspection

Once a class has been defined, there are a number of software tools that can be used to find out about that class. These include `getSlots` that will report the slot names and types, the function `slotNames` that will report only the slot names. These functions are demonstrated using the class A defined above.

```
> getSlots("A")

        s1
"numeric"

> slotNames("A")

[1] "s1"
```

The class itself can be retrieved using `getClass`. The function `extends` can be called with either the name of a single class, or two class names. If called with two class names, it returns `TRUE` if its first argument is a subclass of its second argument. If called with a single class name, it returns the names of all subclasses, including the class itself. However, this is slightly confusing and additional helper functions have been defined in the **RBioinf** package, `superClassNames` and `subClassNames`, to print the names of the superclasses and of the subclasses, respectively. The use of these functions is shown in the code below.

```
> extends("B")

[1] "B" "A"
```

```
> extends("B", "A")

[1] TRUE

> extends("A", "B")

[1] FALSE

> superClassNames("B")

[1] "A"

> subClassNames("A")

[1] "B"
```

These functions also provide information about builtin classes that have been converted via setOldClass.

```
> getClass("matrix")
```

```
No Slots, prototype of class "matrix"

Extends:
Class "array", directly
Class "structure", by class "array", distance 2
Class "vector", by class "array", distance 3, with explicit co
erce

Known Subclasses:
Class "array", directly, with explicit test and coerce
```

```
> extends("matrix")

[1] "matrix"    "array"       "structure" "vector"
```

To determine whether or not a class has been defined, use isClass. You can test whether or not an R object is an instance of an S4 class using isS4. All S4 objects should also return TRUE for is.object, but so will any object with a class attribute.

3.4.1.2 Coercion

The standard mechanism for coercing objects from one class to another is the function **as**, which has two forms. One form is coercion where an instance of one class is *coerced* to the other class, and the second form is an assignment version, where a portion of the object supplied is coerced. The second form is really only applicable to situations where one class is a subclass of the other.

In the example below, we first create an instance of B, then coerce it to be an instance of A. The method for this is automatically available since the classes are nested, and in fact you can also coerce from the superclass to the subclass, with missing slots being filled in from the prototype.

```
> myb = new("B")
> as(myb, "A")

An object of class "A"
Slot "s1":
[1] 0
```

The second form is the assignment form where we replace the A part of `myb` with the new values in `mya`.

```
> mya = new("A", s1 = 20)
> as(myb, "A") <- mya
> myb

An object of class "B"
Slot "s2":
[1] "hi"

Slot "s1":
[1] 20
```

When classes are not nested, the user must provide an explicit version of the coercion function, and optionally of the replacement function. The syntax is `setAs(from, to, def, replace)`, where the `from` and `to` are the names of the classes between which coercion is being defined. The coercion function is supplied as the argument `def` and it must be a function of one argument, an instance of the from class and return an instance of the to class.

In the example below we show the call to `setAs` that defines the coercion between the *graphAM* class, from the **graph** package, and the *matrix* class. The *graphAM* class is a class that represents a graph in terms of an adjacency matrix, so the coercion is quite straightforward. The coercion in the other direction is more complicated.

```
> setAs(from = "graphAM", to = "matrix", function(from) {
    if ("weight" %in% names(edgeDataDefaults(from))) {
        tm <- t(from@adjMat)
        tm[tm != 0] <- unlist(edgeData(from, attr = "weight"))
        m <- t(tm)
    }
    else {
        m <- from@adjMat
    }
    rownames(m) <- colnames(m)
    m
})
```

Calls to `setAs` install a method, constructed from the supplied function, on the generic function `coerce`. You can view the available methods using `showMethods`.

3.4.1.3 Creation of new instances

Once a class has been defined, users will want to create instances of that class. The creation of instances is controlled by three separate but related tools: the specification of a prototype for the class, the creation of an `initialize` method, or through values supplied in the call to `new`. It is essential that the value returned by the initialize method is a valid object of the class being initialized and in general this is a suitably transformed version of the `.Object` parameter. Alternatively, and for complex objects, or large objects, we recommend creating your own constructor function since calls to `new` tend to be somewhat fragile and can be inefficient.

When a call to `new` is made, the following procedure is used. First the class prototype is used to create an initial instance; that prototype is then passed to the `initialize` method hierarchy. Provided any user-supplied `initialize` methods have a call to `callNextMethod`, this hierarchy will be traversed until the default method is encountered. In this method the value is modified according to the arguments supplied to `new` and the result is returned.

The prototype can be set using either a list or a call to `prototype`. In the example below, we define a class, *Ex1*, whose prototype has a random sample of values from the $N(0,1)$ distribution in its `s1` slot.

```
> setClass("Ex1", representation(s1 = "numeric"),
    prototype = prototype(s1 = rnorm(10)))
```

```
[1] "Ex1"

> b = new("Ex1")
> b

An object of class "Ex1"
Slot "s1":
 [1] -1.3730 -0.5483  0.2648  0.0487  1.4423  0.0283  1.1793
 [8] -1.6695 -0.0536  0.0729
```

Exercise 3.6

What happens if you generate a second instance of the Ex1 class? Why might this not be desirable? Examine the prototype for the class and see if you can understand what has happened. Will changing the prototype to `list(s1=quote(rnorm(10)))` *fix the problem?*

When a subclass, such as B from our previous example, is defined, then a prototype is constructed from the prototypes of the superclasses for slots that are not specified in the prototype for the subclass. We see, below, that the prototype for B has a value for the s1 slot, even though none was formally supplied, and that value is the one for the superclass A.

```
> bb = getClass("B")
> bb@prototype

<S4 Type Object>
attr(,"s2")
[1] "hi"
attr(,"s1")
[1] 0
```

If desired, one can define an `initialize` method for a class. The default initialize method takes either named arguments, where the names are those of slots, or one or more unnamed arguments that correspond to instances of any superclass. It is an error to have more than one instance of any superclass or to have the same named argument twice. In constructing the object, the procedure is to first use all values corresponding to superclasses and then the named arguments are applied. Thus, named arguments take precedence.

In the example below, we define two new classes, one a simple class, W, and then a class that is a subclass of both A, defined earlier, and W. When creating new instances of W and A, we made use of named arguments to the initialize method, but when creating a new instance of the WA class, we used the unnamed variant and supplied instances of the superclasses.

```
> setClass("W", representation(c1 = "character"))

[1] "W"

> setClass("WA", contains = (c("A", "W")))

[1] "WA"

> a1 = new("A", s1 = 20)
> w1 = new("W", c1 = "hi")
> new("WA", a1, w1)

An object of class "WA"
Slot "s1":
[1] 20

Slot "c1":
[1] "hi"
```

In the next example we define an initialize method that takes a value for one of the slots and computes the value for the other, depending on the value of the supplied argument. While we named the formal argument to the initialize method b1, that was not necessary and any other name will work. However, we find it helpful to use the slot name if the intention is that the value corresponds to a slot. The user-supplied initialize method overrides the default method, and you can no longer use the slot names, or an instance of a subclass, in the call to new.

```
> setClass("XX", representation(a1 = "numeric",
      b1 = "character"),
      prototype(a1 = 8, b1 = "hi there"))

[1] "XX"

> new("XX")

An object of class "XX"
Slot "a1":
[1] 8

Slot "b1":
[1] "hi there"
```

```
> setMethod("initialize", "XX", function(.Object, ..., b1) {
      callNextMethod(.Object, ..., b1 = b1, a1 = nchar(b1))
  })

[1] "initialize"

> new("XX", b1="yowser")

An object of class "XX"
Slot "a1":
[1] 6

Slot "b1":
[1] "yowser"
```

In our example, it might have been a good idea to include an ... argument so that anyone extending the *XX* class could have some freedom to make use of additional arguments. This comes with the minor cost of confusing your users. If the ... argument is used, and a user supplies a value named a, not knowing that we have supplanted the standard initialize method, they are going to have to work fairly hard to find the problem.

In the code chunk below we establish two classes: *Capital*, which holds only strings in capital letters, and *CountedCapital*, which holds both the string and its length. We then define two initialize methods, one for each class. Important aspects of these methods are the use of ... in the signature to allow for other extensions of the class, and the fact that each method only deals with slots that are specific to its class, leaving the handling of other slots to the classes where they are specified.

```
> setClass("Capital",
          representation=representation(
            string="character"))

[1] "Capital"

> setClass("CountedCapital",
          contains="Capital",
          representation=representation(
            length="numeric"))

[1] "CountedCapital"

> setMethod("initialize",
            "Capital",
```

```
            function(.Object, ..., string=character(0)) {
                string <- toupper(string)
                callNextMethod(.Object, ..., string=string)
            })
```

```
[1] "initialize"
```

```
> setMethod("initialize",
            "CountedCapital",
            function(.Object, ...) {
                .Object <- callNextMethod()
                .Object@length <- nchar(.Object@string)
                .Object
            })
```

```
[1] "initialize"
```

```
> new("Capital", string="MiXeD")
```

```
An object of class "Capital"
Slot "string":
[1] "MIXED"
```

```
> new("CountedCapital", string="MiXeD")
```

```
An object of class "CountedCapital"
Slot "length":
[1] 5
```

```
Slot "string":
[1] "MIXED"
```

```
> new("CountedCapital", string=c("MiXeD", "MeSsaGe"))
```

```
An object of class "CountedCapital"
Slot "length":
[1] 5 7
```

```
Slot "string":
[1] "MIXED"   "MESSAGE"
```

3.4.1.4 Validity

As noted above, validity methods are stored as part of the class definition and can be defined during a call to `setClass` or by calling `setValidity` at some

later time. The validity checking function, if supplied, must return either TRUE
or one or more character strings describing the failures if the object is not a
valid instance of its class. The validity of an object can be tested by calling
validObject. Validity testing is done in essentially a bottom-up manner: first
the validity of all slots are tested and then for each superclass, the validity
method for that class is called, if there is one, and finally the validity method
for the class of the object being tested is invoked.

Validity checking can be problematic. First, it can be expensive and, second,
some transformations will not be atomic with respect to the state of the object,
so premature validity checking will result in a failure. Therefore, by default,
validity is checked when the default initialize method is called; so if there is a
user-defined initialize method that does not call callNextMethod, then validity
will not be checked. Users can call validObject directly.

Serializing and deserializing seem like natural places to test validity, but
this is not currently being done.

Exercise 3.7
*Return to the first representation of the Rectangle class example of Sec-
tion 3.2.3 and write a validity method that ensures that the value placed
in the* area *slot is indeed the product of the width and the height.*

3.4.1.5   Classes without explicit slots

It is possible to define classes without explicit slots. These classes are
defined by providing a prototype but no representation in the call to setClass.
The example below is taken directly from Chambers (1998).

```
> setClass("seq", contains="numeric",
          prototype=prototype(numeric(3)))

[1] "seq"

> s1 = new("seq")
> s1

An object of class "seq"
[1] 0 0 0

> slotNames(s1)

[1] ".Data"
```

Instances of these classes are basically copies of the prototype with a class
attribute. initialize methods can be defined for these classes as is shown
below.

```
> setMethod("initialize", "seq", function(.Object) {
    .Object[1] = 10
    .Object
  })

[1] "initialize"

> new("seq")

An object of class "seq"
[1] 10  0  0
```

Another use for classes with no slots is to define user-controlled extensions of R's internal classes so that methods can be defined for them. It is an error to try and define some methods such as $ and [[on certain builtin classes. But these classes can be trivially extended and then methods can be defined on the extensions. The code below first shows that we fail to attach our method when it is defined for the *integer* class, but that we can set methods on the extended class.

```
> tryCatch(setMethod("[", signature("integer"),
                    function(x, i, j, drop) print("howdy")),
         error = function(e)
         print("we failed"))

[1] "we failed"

> setClass("Myint", representation("integer"))

[1] "Myint"

> setMethod("[", signature("Myint"),
                    function(x, i, j, drop) print("howdy"))

[1] "["

> x = new("Myint", 4:5)
> x[3]

[1] "howdy"
[1] "howdy"
```

In the Bioconductor Project, we initially created metadata packages that consisted of a set of R environments; see Section 2.2.4.3 for more details on this data type. These environments were used as hash tables and access to the data was through functions such as mget and the operators $ and [[. However, as the metadata have grown and become more complex, this approach is no longer tenable and a switch to a lightweight database implementation is underway. For more details on database interfaces, see Chapter 8.4.

The implementation of the database interface has relied on the use of functions to directly access the underlying data tables. But such use is not consistent with using the $ and [[operators. A simple solution is to extend the usual *function* class so that these operators can be used. The code below demonstrates how such an extension can be created.

```
> setClass("DBFunc", "function")

[1] "DBFunc"

> setMethod("$", signature = c("DBFunc", "character"),
      function(x, name) x(name))

[1] "$"
```

Since there is no prototype for this class, we create an instance of the *DBFunc* class by first creating a function in the usual way and then using that value in a call to new to create an instance of the *DBFunc* class.

```
> mytestFun = function(arg) print(arg)
> mtF = new("DBFunc", mytestFun)
> mtF$y

[1] "y"
[1] "y"

> is(mtF, "function")

[1] TRUE
```

An alternative approach is to define a class that contains a function as one of its slots and to then define $ and [[methods for that class. The basic difference between the two approaches is that the one we have used has *is-a* semantics, mtF is a function, while the other approach yields *has-a* semantics. The second approach is needed for any of the reference-like objects in R, such as environments and external pointers.

3.4.2 Types of classes

A class can be instantiable or virtual. Direct instances of virtual classes cannot be created. One can test whether or not a class is virtual using `isVirtualClass`. But note that if the value given is not either the name of an S4 class, or an S4 *classRepresentation* object, this function always returns `TRUE`. Thus, it will often be beneficial to precede this test with a direct ascertainment of whether or not the class is actually an S4 class.

Currently there is some support for sealed classes in S4. One may either seal an S4 class at the time it is created by using the `sealed` argument to `setClass`. Alternatively, the class can be sealed through a call to `sealClass`. A class that is sealed cannot be redefined, and any call to `setClass` will fail when called with the name of a sealed class. Sealing also prevents calls to `setIs` with the sealed class as the first argument. We discuss the semantics and other considerations of `setIs` in Section 3.4.12.2.

3.4.3 Attributes

Due to the way that S4 is currently implemented, attributes should be used with extreme caution. The problem is that basically the S4 system has been implemented via attributes, but without any name mangling, or sequestering of the attributes, so that users can inadvertently make instances non-functional.

In the example below, we examine the attributes on an instance of the *A* class, defined in Section 3.4.1, and see that it has two: one representing the slot `s1` and the other a `class` attribute.

```
> mya = new("A", s1 = 20)
> class(mya)

[1] "A"
attr(,"package")
[1] ".GlobalEnv"

> attributes(mya)

$s1
[1] 20

$class
[1] "A"
attr(,"package")
[1] ".GlobalEnv"
```

And then directly setting that attribute, via `attr`, subverts all of the standard checking for consistency of slots, etc. In the example below, we set the value in the `s1` slot to the letter L, which is not valid. The slot should only hold numeric values, and the object is altered, with no error or warning. Now, one is unlikely to do this intentionally but the chance of it occurring unintentionally should be reduced.

```
> attr(mya, "s1") <- "L"
> mya

An object of class "A"
Slot "s1":
[1] "L"
```

3.4.4   Class unions

An R-level construct that allows for the creation of virtual classes that have a given set of classes as subclasses is quite valuable. In S4, this capability is provided by the function `setClassUnion`. A class may be defined as the union of other classes, that is, as a virtual class defined as a superclass of several other classes. Another way to think of a class union is that the relationship between classes is defined by specifying what the subclasses are. When using `setClass`, the relationship is generally defined by specifying the superclasses. Class unions are useful in method signatures or as slot types in other classes when we want to allow one of several classes to be supplied. As shown below, a fairly common construct is to define a class that allows for either a `list` or `NULL` to be used.

```
> setClassUnion("lorN", c("list", "NULL"))
[1] "lorN"
> subClassNames("lorN")
[1] "list" "NULL"
> superClassNames("lorN")
character(0)
> isVirtualClass("lorN")
[1] TRUE
> isClassUnion("lorN")
[1] TRUE
```

3.4.5   Accessor functions

Accessing slots directly using the @ operator relies on the implementation details of the class, and such access will make it very difficult to change that implementation. In many cases it will be advantageous to provide accessor functions for some, or all, of the components of an object. Suppose that the class *Foo* has a slot named a. To create an accessor function for this slot, we create a generic function named a and a method for instances of the class *Foo*.

```
> setClass("Foo", representation(a = "ANY"))

[1] "Foo"

> setGeneric("a", function(object) standardGeneric("a"))

[1] "a"

> setMethod("a", "Foo", function(object) object@a)

[1] "a"

> b = new("Foo", a = 10)
> a(b)

[1] 10
```

3.4.6   Using S3 classes with S4 classes

S3 classes can be used to describe the contents of a slot in an S4 class, and they can be used for dispatch in S4 methods by first creating an S4 virtualization of the class. This is done with a call to setOldClass, and many such classes are created when the **methods** package is attached.

```
> setOldClass("mymatrix")
> getClass("mymatrix")

Virtual Class

No Slots, prototype of class "S4"

Extends: "oldClass"
```

The resulting S4 classes are virtual classes, so that instances cannot be created directly; instead, you create instances, as for other S3 classes, by directly manipulating the `class` attribute. S3 instances will be dispatched on correctly and can be used to populate the slots of an S4 object that uses them. All classes created by a call to `setOldClass` inherit from the class *oldClass*.

```
> setClass("myS4mat", representation(m = "mymatrix"))

[1] "myS4mat"

> x = matrix(1:10, nc = 2)
> class(x) = "mymatrix"
> m4 = new("myS4mat", m = x)
```

The set of all exposed S3 classes that have been converted to S4 classes in the **methods** package can be obtained by using the fact that they all inherit from the *oldClass* class. One might expect this to find all such classes, but unfortunately classes defined in other packages do not register with the class definition, so they need to be searched for via other methods.

```
> head(subClassNames(getClass("oldClass")))

[1] "data.frame"    "factor"        "table"
[4] "summary.table" "lm"            "POSIXt"
```

Exercise 3.8
Write a function that searches every package on the search path for any class that extends oldClass.

3.4.7 S4 generic functions and methods

Generic functions are created by calls to `setGeneric` and, once created, methods can be associated with them through calls to `setMethod`. The arguments of the method must conform, to some extent, with those of the generic function. The method definition indicates the class of each of the formal arguments and this is called the signature of the method. There can be, at most, one method with any signature.

In most cases the call to `setGeneric` will follow a very simple pattern. There are a number of arguments that can be specified when calling `setGeneric` and we begin by describing the first two: the `name` argument specifies the name of the generic function while the `def` argument provides the definition for the generic function. In almost all cases the body of the function supplied as the `def` argument will be a call to `standardGeneric` since this function is used to both dispatch to methods based on the supplied arguments to the

generic function and it also establishes a default method that will be used if
no function with matching signature is found.

The syntax is quite straightforward. The `def` argument is a function, each
named argument can be dispatched on, and the ... argument should be used
if other arguments to the generic will be permitted. These arguments cannot
be dispatched on, however. So in the code below, the generic function has two
named arguments, `object` and `x`, and methods can be defined that indicate
different signatures for these two arguments.

```
> setGeneric("foo", function(object, x) standardGeneric("foo"))

[1] "foo"

> setMethod("foo", signature("numeric", "character"),
      function(object, x) print("Hi, I'm method one"))

[1] "foo"
```

Exercise 3.9
*Define another method for the generic function foo defined above, with a
different signature. Test that the correct method is dispatched to for different
arguments.*

Any argument passed through the ... argument cannot be dispatched on.
It is possible to have named arguments that are not part of the signature of
the generic function. This is achieved by explicitly stating the signature for
the generic function using the `signature` argument in the call to `setGeneric`,
as is demonstrated below. In that case it may make sense for a method to
provide default values for the arguments not in the signature.

```
> setGeneric("genSig", signature = c("x"), function(x,
      y = 1) standardGeneric("genSig"))

[1] "genSig"

> setMethod("genSig", signature("numeric"), function(x,
      y = 20) print(y))

[1] "genSig"

> genSig(10)

[1] 20
```

Unlike S3, where dispatch using `UseMethod` does not return, in S4 control will return to the generic function so post-processing is possible. The code below gives a simple example demonstrating that control has returned.

```
> setGeneric("foo", function(x, y, ...) {
      y = standardGeneric("foo")
      print("I'm back")
      y
 })

[1] "foo"

> setMethod("foo", "numeric", function(x, y, ...) {
      print("I'm gone")
 })

[1] "foo"

> foo(1)

[1] "I'm gone"
[1] "I'm back"
[1] "I'm gone"
```

Whether or not a function is a generic function can be determined using `isGeneric`. Generic functions can be removed using `removeGeneric`, but this is not too useful since only generic functions defined in the user's workspace are easily removed.

We want to dispel a prevalent misconception about generic functions, or any other R objects for that matter. There is a belief that for any given name (such as `plot`, for example), there can be only one generic function. This is not true, and generic functions are no different than any other function. Every package can define its own generic function `foo` and there is no need for the argument lists to agree in any way. When a call to `foo` is evaluated, which generic function is used is determined by the usual scoping rules; see Section 2.12.1 for more details. And when a method is defined and associated with a generic function, using a call to `setMethod`, for example, the programmer must be careful to ensure that the method is associated with the intended generic function.

To find all generic functions that are defined, and the packages that they are defined in, use the function `getGenerics`, with no arguments. This function relies on data in a table stored in the methods package. If `getGenerics` is called with an argument that corresponds to a package, then it will list all generic functions for which there is a method defined in the package, not just the generic functions defined in that package. In the example below, we load the **Biobase** package and then try to find all generic functions that are defined in it.

```
> library("Biobase")
> allG = getGenerics()
> allGs = split(allG@.Data, allG@package)
> allGBB = allGs[["Biobase"]]
> length(allGBB)

[1] 78
```

Next we use the `where` argument to only get generic functions defined in **Biobase**. But we see that there are more generic functions reported than above. This is because in using this approach, we are getting all generic functions that have a method defined for them in the package, not all generic functions defined in the package. If we restrict these generic functions to those whose package description is `Biobase`, then we get the same answer as above.

```
> allGbb = getGenerics("package:Biobase")
> length(allGbb)

[1] 90
```

3.4.7.1   Evaluation model for generic functions

When the generic function is invoked, the supplied arguments are matched to the arguments of the generic function; those that correspond to arguments in the signature of the generic are **evaluated**. This eager evaluation of arguments in the signature is a substantial change from the lazy evaluation semantics that are used for standard function invocation.

Once evaluation of the generic function begins, all methods registered with the generic function are inspected and the applicable methods are determined. A method is applicable if for all arguments in its signature, the class specified in the method either matches the class of the supplied argument or is a superclass of the class of the supplied argument. The applicable methods are ordered from most specific to least specific. Dispatch is entirely determined by the signature and the registered methods at the time evaluation of the generic function begins.

3.4.8   The syntax of method declaration

Methods are declared and assigned to generic functions through calls to `setMethod`. They can be removed through a call to either `removeMethod` or `removeMethods`. The method should have one argument matching each argument in the signature of the generic function. These arguments can correspond to any defined class or they can be either of the two special classes *ANY* and *missing*. Use *ANY* if the method will accept any value for that argument. The class *missing* is appropriate when the method will handle some, but not all, of the arguments in the signature of the generic.

Exercise 3.10
Write different methods for the generic function foo defined above, that make use of ANY, and missing in the signature. Test these methods to be sure they behave as you expect.

When ... is an argument to the generic function, you can define methods with named arguments that will be handled by the ... argument to the generic function. But some care is needed because these arguments, in some sense, do not count. There can be only one method, with any given signature (set of classes defined for the formal arguments to the generic), regardless of whether or not other argument names match.

```
> setGeneric("bar", function(x, y, ...) standardGeneric("bar"))

[1] "bar"

> setMethod("bar", signature("numeric", "numeric"),
      function(x, y, d) print("Method1"))

[1] "bar"

> ##removes the method above
> setMethod("bar", signature("numeric", "numeric"),
      function(x, y, z) print("Method2"))

[1] "bar"

> bar(1,1,z=20)

[1] "Method2"

> bar(2,2,30)

[1] "Method2"

> tryCatch(bar(2,4,d=20), error=function(e)
            print("no method1"))

[1] "no method1"
```

If `setMethod` is called on a function for which there is no corresponding S4 generic function, one is created automatically and the existing function is established as the default method for that generic function.

3.4.9   The semantics of method invocation

When a generic function is invoked, the classes of all supplied arguments that are in the signature of the generic function form the *target signature*. A method is said to *be applicable* for this target signature if for every argument in the signature the class specified by the method is the same as the class of the corresponding supplied argument, a superclass of that class, or has class `ANY`. To order the applicable methods, we need a metric on the classes. And a simple one is that if the classes are the same, the distance is zero; if the class in the signature of the method is a direct superclass of the class of the supplied argument, then the distance is one, and so on. The distance from a class to `ANY` is chosen to be larger than any other distance. The distance between an applicable method and the target signature can then be computed by summing up the distances over all arguments in the signature of the generic function, and these distances can then be used to order the methods.

Once the the ordered list of methods has been computed, control is passed to the most specific method. Evaluation is essentially the same as for any function, except that the formal arguments to the generic have already been evaluated. Evaluation of the body of the method is carried out in essentially the same way as the evaluation of any ordinary function. But, if the body of a method contains a call to `callNextMethod`, then control is passed to the next method in the linearization that was computed by the generic function. The arguments to the generic are rematched (but not reevaluated) to the next method, and the body of that method is evaluated. Control will return to the calling method, so a more specific method can choose to either perform preprocessing or post-processing.

The next method is invoked with the same set of arguments as the current method, but with the values of those arguments being the values that correspond to their values in the current method. Named arguments to the generic are only evaluated once, at the time the generic function is invoked. Any argument that is *missing* in the current call is missing in the next method. The effect is essentially that the evaluation of the next method uses the current evaluation environment for bindings to all formal arguments. Other symbols in the current evaluation argument are not available.

Methods are lexically scoped. That means if the function, or closure, that is used for the method, was created in such a way as to have an enclosing environment (see Section 2.13 for a more detailed description of closures and lexical scope in R), then that information is retained with the method.

3.4.10 Replacement methods

Replacement functions were discussed in Section 2.7, and S3 replacement functions were discussed in Section 3.3.6. S4 replacement methods are quite similar. They require that an appropriate S4 generic function be defined; usually its name is of the form `genfun<-`. You must ensure that the method returns the whole object and that the last argument is named `value`. This ensures that R can always identify the value that is going to be assigned.

We continue the example given above and define a method that will change the value in the slot named a. We first define an appropriate generic function and then use `setReplaceMethod` to define the replacement method.

```
> setGeneric("a<-", function(x, value) standardGeneric("a<-"))

[1] "a<-"

> setReplaceMethod("a", "Foo", function(x, value) {
      x@a = value
      x
  })

[1] "a<-"

> a(b) = 32
> b

An object of class "Foo"
Slot "a":
[1] 32
```

3.4.11 Finding methods

One of the strengths of R is the ability to program on the language. In order to do that, we need to be able to do more than simply determine whether a generic function exists. We will often need to be able to determine which methods are registered with a particular generic function. At other times we will want to be able to determine whether a particular signature will be handled by a generic. Functionality of this sort is provided by the functions listed next. In all cases there are more parameters and details than can be presented here, so interested readers are referred to the online manual for a more comprehensive treatment.

showMethods shows the methods for one or more generic functions. The class argument can be used to ask for all methods that have a particular class in their signature. The output is printed to `stdout` by default and cannot easily be captured for programmatic use.

getMethod returns the method for a specific generic function whose signature is congruent with the specified signature. An error is thrown if no such method exists.

findMethod returns the packages in the search path that contain a definition for the generic and signature specified.

selectMethod returns the method for a specific generic function and signature, but differs from `getMethod` in that inheritance is used to identify a method.

existsMethod tests for a method with a congruent signature (to that provided) registered with the specified generic function. No *inheritance* is used. Returns either `TRUE` or `FALSE`.

hasMethod tests for a method with a congruent signature for the specified generic function. It seems that this would always return `TRUE` (since there must be a default method). It does return `FALSE` if there is no generic function, but it seems that there are better ways to handle that.

3.4.12   Advanced topics

3.4.12.1   Setting methods on $

The $ operator is special in R and, as discussed in Section 2.5, this operator does not evaluate its second argument. But S4 generic functions evaluate all arguments in their signature, so this argument cannot be in the signature and cannot be dispatched on. But further, if the method intends to use the $ operator, then some effort is needed to construct the call. In the example below we show how this was achieved in the **Biobase** package.

```
> setMethod("$", "eSet", function(x, name) {
      eval(substitute(phenoData(x)$NAME_ARG,
                   list(NAME_ARG = name)))
  })
```

3.4.12.2   setIs

The S4 system also allows users to establish an inheritance relationship between two classes even when there is not a direct inclusion. This is accomplished through the `setIs` function. The relationships that can be described by `setIs` have the potential to be problematic,and one should use this function with caution. Perhaps the only somewhat innocuous use of `setIs` is to add classes to a class union. More details and examples on its use are given in Chambers (2008).

3.4.12.3 Dispatching on ...

Recently the issue of whether to dispatch on the ... argument has been raised. The justification for such dispatch is that for some functions, such as c, there is a reasonable model for dispatching on a collection of values supplied via the ... argument. One implementation, proposed by L. Tierney, is shown in the code below. Dispatch is handled by recursively dividing the values supplied in the ... formulation into sets of two, and using a helper function with two named arguments that can be dispatched on.

```
> cnew = function(x, ...) {
    if (nargs() < 3)
        c2(x, ...)
    else c2(x, cnew(...))
}
```

We have defined a new function, cnew, rather than use c so as not to interfere with regular dispatch.

```
> setGeneric("c2", function(x, y) standardGeneric("c2"))

[1] "c2"
```

And methods can be written for c2. So, for example we could write a method to add numeric values, as is shown below. Other methods could be written to deal with other types of data.

```
> setMethod("c2", signature("numeric", "numeric"),
        function(x, y) x + y)

[1] "c2"

> cnew(1, 2, 3, 4)

[1] 10
```

3.5   Using classes and methods in packages

An area where there is not yet consensus on what should happen and how is the problem of having a package provide a method for a generic function that is defined elsewhere. If the generic exists in a known package then things are straightforward. The `where` argument in the call to `setMethod` should be used to ensure that the appropriate generic is used. Recall that there can be multiple generic functions with the same names, so it is your responsibility to ensure that the method is attached to the correct one.

Things are less clear if the method is to be attached to a non-generic function, say one from the **base** package. Because then, you will not be able to easily tell where some other package has already created a local generic. You can search for a generic function with the name you want, but it is not easy to be sure that is defined for the function you want to use. If a package defines a method for a generic that exists in another package, then the association of that method with the appropriate generic function must occur at package load time. If these computations are carried out at package build time, the net effect is to create a new generic function within the package and register the method with it and dispatch will not occur as intended.

3.6   Documentation

3.6.1   Finding documentation

Either a direct call to `help` or the use of the `?` operator will obtain the help page for most functions. To find out about classes an infix syntax is used, where the word `class` precedes the question mark. The syntax for displaying the help page for the *graph* class, from the **graph** package is shown below.

```
class?graph
help("graph-class")
```

Help for generic functions requires no special syntax; one just looks for help on the name of the generic function. Finding help for methods is less easy. There several syntactic variants, but none are completely satisfactory. We show the syntax for two different ways to find the help page for a method for the `nodes` generic function, for an argument of class *graphNEL*.

```
method?nodes("graphNEL")
help("nodes,graphNEL-method")
```

For the **RBioinf** package, we have developed a function that provides a different, and hopefully easier to use, interface to the help system. The function is called `S4Help`, and currently takes the name of either a S4 generic function or a S4 class and provides a selection menu to choose a help page. If the supplied name is a class, then that class and any superclass can be selected. If the supplied name corresponds to a generic function, then that function, or any of its methods, can be selected. See the help page for `S4Help` for more details.

3.6.2 Writing documentation

Documenting S4 classes and methods is quite similar to documenting other R objects, but there are some important differences; many of them are detailed in Writing R Extensions (R Development Core Team, 2007c). Here we outline some of the basic concepts. The current state of formal organization for documenting S4 classes and methods is relatively incomplete and we consider some extensions and improvements as well. There are two functions that provide shell documentation: `promptClass` for classes and `promptMethods` for the methods of a supplied generic function.

Documentation of a class should require the specification of all of the arguments that can be supplied to `setClass`.

Generic functions are just like any other functions and should be documented as such. It would be nice if there was some automatic way to integrate the documentation of methods with that of the corresponding generic functions when packages are attached and detached, but that is not possible with R's current documentation system.

For packages, the author could document both the generic and all defined methods on a single manual page. Method documentation should always link to the corresponding manual page for the generic. It should make it clear which arguments have been specialized and describe the manner in which this specialization has affected computations and return values, if at all. For any specialized arguments, the manual page for the method should link to the appropriate class documentation pages.

3.7 Debugging

While general debugging is discussed in Chapter 9, we provide some specific advice on how to debug S4 methods. First, it is not possible to use `debug` directly because the methods are not available in a form that allows users to easily request that the method be debugged. One can `debug` the generic function but that is typically not very satisfactory, and there is no easy way to step into the method that is dispatched to.

Instead, the function `trace` can be used to debug S4 methods. This function is discussed in more detail in Section 9.3.5 and so here we will simply give the syntax for debugging a particular method. In the code below, the first command shows how to begin debugging on entry into the `dim` method with signature `eSet`. The second command also places a call to `browser` on exit from the method.

```
trace("dim", browser, signature = c("eSet"))
trace("dim", browser, exit=browser, signature = c("eSet"))
untrace("dim", signature = c("eSet"))
```

3.8   Managing S3 and S4 together

Perhaps one of the more unfortunate aspects of OOP in R is that users are left to manage rather a lot of the interface between S3 and S4. Here we describe some of the tools that can be used to help detect and work around issues that might arise.

Testing for inheritance is done differently between S3 and S4. The former uses the function `inherits` while the latter uses `is`. The unfortunate part is that both `inherits` and `is` give partial answers (not errors) if applied to instances from the other class system. In the example below, x is an S3 instance, so `inherits` does correctly indicate the inheritance relationship but `is` does not.

```
> x = 1
> class(x) = c("C1", "C2")
> is(x, "C2")

[1] TRUE

> inherits(x, "C2")

[1] TRUE
```

Exercise 3.11
Show that for S4 classes, `is` gets the inheritance correctly while `inherits` does not.

Now one can make use of `setOldClass` to basically tell S4 what the class relationships should be. And if this is done, then `is` is indeed able to correctly identify the inheritance relationships.

```
> setOldClass(c("C1", "C2"))
> is(x, "C2")

[1] TRUE
```

The function `isS4` returns `TRUE` for an instance of an S4 class. For primitive functions that support dispatch, S4 methods are restricted to S4 objects. The function `asS4` can be used to allow an instance of an S3 class to be passed to an S4 method.

In the next example we show that when x is an S3 instance, we do not dispatch to the S4 method, but once we use `asS4`, then dispatch to the S4 method occurs.

```
> x = 1
> setClass("A", representation(s1 = "numeric"))

[1] "A"

> setMethod("+", c("A", "A"), function(e1, e2) print("howdy"))

[1] "+"

> class(x) = "A"
> x + x

[1] 2
attr(,"class")
[1] "A"

> asS4(x) + x

[1] "howdy"
[1] "howdy"
```

3.8.1 Getting and setting the `class` attribute

Another difference between the S3 and S4 systems comes from the return value for the `class` function. For instances of S3 classes, the `class` attribute should hold the names of all classes that the object inherits from and this vector is returned. For instances of S4 objects, the `class` attribute is always of length one, the most specific class, and this is returned. Inheritance is determined from the existing class definitions. Use of the `oldClass` mechanism

muddies the water somewhat as the instances are S3, but they need only have
length one class attributes since all inheritance can be determined from the S4
class definitions that are created as a result of setting up the `oldClass`. The
basic message is that the `class` function is only reliable for finding the most
specific class of an instance. To find out about inheritance, you should use `is`
for instances of S4 classes, including those S3 classes that have an `oldClass`
specification. You should use `inherits` for all other instances of S3 classes.

One place where the paradigm described above might fail is if an S4 method
dispatches on a class that is a subclass of class for which there is an S3
method. Then the method for the subclass will be preferred over the method
for the superclass, and that is not what should happen. In that case, you
have little choice but to translate the S3 method into an S4 method. If the S3
method does not rely on any of the S3 dispatch mechanisms such as variables
like `.Generic`, and it has no calls to `NextMethod`, then this can be done quite
simply. One need only call `setMethod` with the appropriate signature and
the S3 method as an argument. Chambers (2008) suggests that the explicit
calling of the S3 method is preferred in some settings and that is an alternative.
Pseudo-code for these two cases is shown below.

```
> setMethod("foo", "myclass", myS3Method)
> setMethod("foo", "myclass", function(x, y, ...) myS3Method(x,
    y, ...))
```

3.8.2 Mixing S3 and S4 methods

Having a generic function that can dispatch to either S3 or S4 methods
is reasonably straightforward. This is achieved using an S4 generic function
with a default method that contains a call to the S3 function `UseMethod`. If
there is an existing S3 generic function, then calling `setGeneric` with its name
as the argument will create an S4 generic with the existing S3 generic as its
default method. Dispatch is then first carried out for S4 and if no method is
found, then the default method is reached and S3 dispatch begins.

In the example below, we create an S3 generic for a simple class, then create
and S4 generic, and finally show that the S3 generic is indeed the default
method for the S4 generic.

```
> testG = function(x, ...) UseMethod("testG")
> setGeneric("testG")

[1] "testG"

> getMethod("testG", signature = "ANY")
```

```
Method Definition (Class "derivedDefaultMethod"):

function (x, ...)
UseMethod("testG")

Signatures:

target
defined
```

3.9 Navigating the class and method hierarchy

We now discuss the tools that are available for navigating and understanding the class and method hierarchy. The main use case is that of trying to understand how the classes in a particular package are associated with each other, and to understand the method hierarchy of a given generic function. We have written some tools that are supplied in the **RBioinf** package to carry out these tasks. We will use the **graph** package as our example and rely on functions from both the **RBGL** package and the **Rgraphviz** package to manipulate and render the resulting graphs. To obtain all of the classes that are defined in a particular package, use `getClasses`, which can be either given the position of the package in the search path or the name space. The former gives all exported classes while the latter should give all classes, whether they are exported or not.

```
> graphClasses = getClasses("package:graph")
> head(graphClasses)

[1] "attrData"    "bzfile"      "clusterGraph"
[4] "connection"  "distGraph"   "edgeSet"
```

We would then like to get a better sense of how they are interrelated and what the complete class hierarchy is. We can most easily do that by visualizing the class hierarchy, and to do that we construct a graph based on all of the classes in the **graph** package.

```
> graphClassgraph = classList2Graph(graphClasses)
```

Once we have the graph, we can interrogate it and then render some of the interesting parts of it using **Rgraphviz**. We first find out how many connected components there are, and then examine how big each is.

```
> ccomp = connectedComp(graphClassgraph)
> complens = sapply(ccomp, length)
> length(ccomp)

[1] 7

> table(complens)

complens
1 3 5 6
4 1 1 1
```

There are four components of size 1; that means that they are classes with no subclasses and no superclasses. We can print their names.

```
> unlist(ccomp[complens == 1], use.names = FALSE)

[1] "graph:attrData"   "graph:multiGraph" "graph:renderInfo"
[4] "graph:simpleEdge"
```

Next we might want to plot the larger components to see what the set of *is-a* relationships are. In the code below we select the largest connected component and create a subgraph that contains only those nodes. When rendering, we set the shape of the nodes to be ellipses so that the text is more easily read.

```
> subGnodes = ccomp[[which.max(complens)]]
> subG = subGraph(subGnodes, graphClassgraph)
> nodeRenderInfo(subG) <- list(shape="ellipse")
> attrs = list(node=list(fixedsize = FALSE))
> x = layoutGraph(subG, attrs = attrs)
> renderGraph(x)
```

The largest is-a hierarchy arises from a set of S3 classes for connections that was extended in order to allow dispatching on it.

Exercise 3.12
Plot the graph that corresponds to the second largest connected component. What classes does it contain?

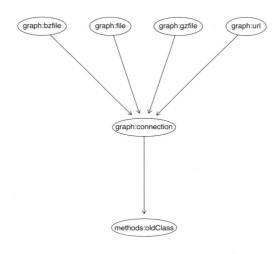

FIGURE 3.1: The graph of all subclass and superclass relationships in the **graph** package.

The inheritance graph, or set of is-a relationships, is only part of the story. We will also need to examine the *has-a* relationships to get a complete view of the class hierarchy. In this case we are interested in the graph that again uses all classes to define the nodes, but a directed edge is drawn from class A to class B if A has a slot that contains an instance of class B.

Exercise 3.13
Write a function to compute the has-a relationships between all classes in a package. You will probably want to also include classes that are not defined in the package, but appear in slot specifications. You might not want to worry too much about `classUnions` *at this point, but a comprehensive solution would need to deal with them.*

Chapter 4

Input and Output in R

4.1 Introduction

Reading and writing data, either on the local computer or over the Internet, is often an important part of a computational task. In this chapter we discuss the different ways in which R interacts with the file system and other external resources. There are a number of functions specifically oriented toward reading and writing from files, some specifically designed for reading formatted data and others for a variety of other interactions. There is some fragmentation and redundancy, and almost all tasks can be carried out using connections.

There is substantial material provided in the R Data Import/Export manual (R Development Core Team, 2007a) that comes with every copy of R. You may want to consult that reference and the help files for some of the more gory details. Among the topics covered there, and not here, are the importing of data from other software systems such as SAS. Discussion of the use of XML and access to relational databases is deferred to Chapter 8.

The notion of a file system, and accessing it, has become somewhat more general in recent times. We are seldom ever restricted to only the system on the local machine since almost all computer users, regardless of operating system, use some form of network file system that shares files across a variety of different computers. Further, the advent of the Internet and the notions of URLs and URIs have led to the use of web addresses as reasonable surrogates for local file names. And finally, debates on whether or not there is a real difference between a file system and a database continue. In many ways this is quite useful, since it helps programmers realize the commonalities between these external types of storage and internal storage. By internal storage, we mean storage, or memory that is under the direct control of the program or system being used; in our case, R.

Users of R and Bioconductor will want to read data from different sources, combine them and process them. This procedure will often result in the generation of intermediate data resources that may be stored in a database, written to a file system or stored in R's internal format. Loading packages, see Chapter 7, requires the reading and processing of files from the file system, while downloading and installing packages requires Internet access and the resolution of Internet addresses. When users quit from R and want to save their intermediate results, a file is written on the local file system. Somewhat less obviously, interactions with the terminal, where R commands are typed as input and various answers and values printed as output, can be treated in much the same way; see Section 4.5 for more details.

4.2   Basic file handling

R provides many file handling capabilities. Reading files on the local system requires knowing either the relative or absolute path to those files. We distinguish between files and directories; a directory is also referred to as a folder on some systems. Most file systems are organized with a specific hierarchy of directories, each of which may contain files and other directories. When R is running, there is always a working directory and relative paths are defined with respect to the working directory. The function `getwd` returns a string with the path to the current working directory. This is the directory that R will use as a default location to read or write files. On Unix systems the current working directory is the directory R was invoked from, while on Windows the initial working directory is the location of the R binary. The current working directory can be changed by a call to `setwd` with the path to the new directory, either absolute or relative, as the sole argument. An error is signaled if the new directory is inaccessible or nonexistent, while the current directory is returned, invisibly, if the change in working directory is successful. Any function that changes the current working directory should reset it to the location it had on entry into the function. In the example below, we show how to access the current working directory using the `getwd` function.

```
> getwd()

[1] "/Users/robert/RG/Lab/RBioinf"
```

There are times when it will be necessary to access files that have been provided with R or with one of the add-on packages that are installed. The `R.home` function returns the home directory for the version of R that is currently running. This is the top-level installation directory of R, and system files provided with R can be accessed using this as the starting point. For files in R packages or files supplied with R, the `system.file` function should be used to obtain the appropriate path.

Exercise 4.1
*What is the location reported by `R.home` on your system? What is the path to the **stats** package?*

Once the appropriate name, including the path, has been obtained, the next step is to open the file for reading or writing, or both. Opening a file, from within R, is handled by the function `file`. Syntax of the form `file("test1", open="rw")` will open the file named `test1` for both reading and writing. In R there are a fairly small number of file handles available so it is important that when a file is no longer needed, its file handle be closed, typically using a call to `close`. It is often also useful to use functionality described in Section 2.12.5 to ensure that files and other file system resources opened during a function call are closed, regardless of how the function is exited.

The simplest function for reading data is `readLines`, which reads lines of text from a connection into a character vector. No processing of the lines is done, so the user has complete control over how the input data are parsed and interpreted. In the code shown below, we open a file named `test1`, which is supplied with the **RBioinf** package, and read its contents.

```
> fp = system.file("extdata/test1", package = "RBioinf")
> f1 = file(fp, open = "r")
> readLines(f1)

[1] "asdf"  "adf"    "ss,bb"

> close(f1)
```

Exercise 4.2
What does `system.file` return if there is no file with the specified name? How many lines were in the file `test1`?

A slightly more user-friendly interface is provided by the function `scan`, which accepts the name of a file or a connection and reads white-space separated values. `scan` has many options that allow fairly sophisticated control of the reading, including `skip` for skipping initial lines in the file, `comment` to indicate a comment character, and `what` to control the types of the return values. `scan` can also be used at the command line, interactively to read in data from the keyboard. On the other hand, `readLines` allows for line-by-line processing, which is very similar to the way in which Perl processes files.

```
> scan(fp, what = "")

[1] "asdf"  "adf"    "ss,bb"
```

One of the major problems that arises when writing platform-independent software, such as R packages, is the fact that many aspects of the file systems are often quite different. For example, whether or not the file system is case sensitive, or the file separator; some systems use a forward slash while others use a backward slash. On any system, the file separator can be found by accessing .Platform$file.sep. Since opening files is an important part of many different tasks, there is a system-independent way of specifying the path to a file. The routine file.path accepts a set of comma-separated inputs and concatenates them using the value of the fsep argument, which by default is set to .Platform$file.sep. This is the safest way to construct paths to files. The path to files in R packages can be found using system.file.

```
> file.path(R.home(), "doc")

[1] "/Users/robert/R/R27/doc"

> system.file(package = "RBioinf")

[1] "/Users/robert/R/R27/library/RBioinf"
```

Files in any directory can be listed using the list.files function. By default, the current working directory is used, but any other directory can be specified using the path argument. The dir function is an alias for list.files and can be used in exactly the same way.

list.files has the following formal arguments:

path a character vector of full path names; the default is the current working directory.

pattern an optional regular expression. Only file names that match the regular expression will be returned. The default is NULL, which matches everything.

all.files a logical value. If FALSE, only the names of visible files are returned. If TRUE, all file names will be returned. The default is FALSE.

full.names a logical value. If TRUE, the directory path is prepended to the file names. If FALSE, only the file names are returned. The default is FALSE.

In the code chunk below, we demonstrate the use of a few of the options available for the list.files function. First, we set the path to the top-level installation directory for R (you should see the same outputs as are listed here if you run these same commands on your computer). Notice that we obtain the current working directory from the initial call to setwd and then restore it in one of the later code chunks.

```
> cd = setwd(R.home())
> list.files(path = "doc")

 [1] "AUTHORS"        "COPYING"          "COPYING.LIB"
 [4] "COPYRIGHTS"     "CRAN_mirrors.csv" "FAQ"
 [7] "KEYWORDS"       "KEYWORDS.db"      "Makefile"
[10] "Makefile.in"    "R.1"              "R.aux"
[13] "RESOURCES"      "Rscript.1"        "THANKS"
[16] "html"           "manual"

> list.files(pattern = "Make")

[1] "Makeconf"       "Makeconf.in"    "Makefile"
[4] "Makefile.bak"   "Makefile.in"    "Makefrag.cc"
[7] "Makefrag.cc_lo" "Makefrag.cxx"   "Makefrag.m"

> list.files(pattern = "Make", full.names = TRUE)

[1] "./Makeconf"       "./Makeconf.in"    "./Makefile"
[4] "./Makefile.bak"   "./Makefile.in"    "./Makefrag.cc"
[7] "./Makefrag.cc_lo" "./Makefrag.cxx"   "./Makefrag.m"
```

Information about files, such as their size, modification date, etc., is provided by the `file.info` function. This function can be used to distinguish a file from a directory. In the code below, some of the `file.info` capabilities are demonstrated. First, we make use of `setwd` to change the working directory. Since `setwd` accepts relative paths, the net effect of the first command is to change the working directory to the `doc` subdirectory of the R home directory (we set the R home directory as the active directory in the code chunk above).

```
> setwd("doc")
> getwd()

[1] "/Users/robert/R/R27/doc"

> file.info("KEYWORDS")$isdir

[1] FALSE

> file.info("manual")$isdir

[1] TRUE
```

You can put this together with some of the commands above to create sets of files or directories that can be used for other tasks. In the example below, we obtain a list of files in the current directory and then remove the directories from that list. The variable `files` contains the names of the files; all directories have been excluded.

```
> x = list.files()
> x

 [1] "AUTHORS"        "COPYING"          "COPYING.LIB"
 [4] "COPYRIGHTS"     "CRAN_mirrors.csv" "FAQ"
 [7] "KEYWORDS"       "KEYWORDS.db"      "Makefile"
[10] "Makefile.in"    "R.1"              "R.aux"
[13] "RESOURCES"      "Rscript.1"        "THANKS"
[16] "html"           "manual"

> files = x[!file.info(x)$isdir]
> files

 [1] "AUTHORS"        "COPYING"          "COPYING.LIB"
 [4] "COPYRIGHTS"     "CRAN_mirrors.csv" "FAQ"
 [7] "KEYWORDS"       "KEYWORDS.db"      "Makefile"
[10] "Makefile.in"    "R.1"              "R.aux"
[13] "RESOURCES"      "Rscript.1"        "THANKS"

> setwd(cd)
```

In other cases you might want to know whether you have permission to read, write or execute particular files. This information can be obtained by using the `file.access` function. Be careful though, as it has non-standard return values, 0 for success and -1 for failure. While one might be tempted to use this function to test for the ability to access a file, there are some reasons why that will not always work, and either `try` or `tryCatch` (Section 2.11) should be used to gracefully deal with a failure to open a named file.

Exercise 4.3
Find the location of the library directory for your version of R. How many directories are there? How many plain files? How many directories can you execute?

4.2.1   Viewing files

There are many different strategies that can be used to view the contents of a file from within R. The function `file.show` uses the same tools that the help system uses to display the file in the R console. Alternatively, one could use the `system` function to directly access a command such as `cat` or `more`. The following simple method extends the `head` function to support external files.

```
> head.file = function(x, n = 6, ...) readLines(x,
+     n)
```

Exercise 4.4
Write a similar function to implement `tail` functionality for files in R.

4.2.2 File manipulation

R maintains a per-session temporary directory for use, and the path to this directory is found using `tempdir`. By default, `tempfile` returns a file name, which is the complete path for a file within the temporary directory. Note that `tempfile` does not actually create the file; it merely provides an available file name. There have been reports of collisions, and users who want genuinely unique file names should consider using the `Ruuid` package to generate a name that is guaranteed to be unique.

In the code below, we first use the function `tempdir` to obtain the location of a per-session temporary directory that the user has permission to write to. This temporary directory is deleted when the R session is terminated and so is not a suitable location for files that are intended to be used after the R session is ended. The two calls to `tempfile` generate file names within the directory identified by `tempdir`. The call to `tempfile` produces the name of the file but does not create a file on the local file system.

```
> tempdir()

[1] "/tmp/RtmpHPmzRh"

> tmp1 = tempfile()
> tmp2 = tempfile()
> tmp1

[1] "/tmp/RtmpHPmzRh/filebc5816b"

> tmp2

[1] "/tmp/RtmpHPmzRh/file353f7788"
```

Exercise 4.5
Verify the claim that the calls to `tempfile` *do not create files.*

Now we will demonstrate a series of file manipulation functions, all of which are fairly self-explanatory. First, using `file.create`, we create the file /tmp/RtmpHPmzRh/filebc5816b; then we test to see if it exists using `file.exists`. We next test whether we can write to it using `file.access` and then remove the file using `file.remove`; and finally we test, again, for its existence. All file manipulation functions expand path names using `path.expand` and take a vector of file names whose values are operated on simultaneously. In general, the functions return logical values, either `TRUE` or `FALSE`, indicating whether they succeeded or failed, with `file.access` being an exception to that rule.

Some caution in using `file.create` is needed, since it will either create the file, if it does not exist, or truncate it (i.e., empty it) if it does; thus, you can easily and inadvertently remove a file. We first use `tempfile` to get a name for a temporary file and then show how to create it, test for its existence and remove it.

```
> tmp1 = tempfile()
> file.create(tmp1)

[1] TRUE

> file.exists(tmp1)

[1] TRUE

> file.access(tmp1, 2)

/tmp/RtmpHPmzRh/file21980f48
                           0

> file.remove(tmp1)

[1] TRUE

> file.exists(tmp1)

[1] FALSE
```

As shown in the next code chunk, the function `path.expand` will expand the tilde, $\sim$, on platforms that support it. The functionality is somewhat more limited than the expansion in the shell; only paths relative to the user's home directory will be expanded. Finding the path to your home directory is done in a somewhat peculiar fashion by using only the tilde as the argument.

In the code below, we show how to use `path.expand` to find the location of my home directory and then create a variable with the path to my R executable using `file.path`.

```
> myhome = path.expand("~")
> myhome

[1] "/Users/robert"

> toR = file.path(myhome, "bin", "R")
> toR

[1] "/Users/robert/bin/R"
```

There are other path manipulation functions beyond `file.path` that can be useful. The function `basename` returns the substring of the supplied argument that appears after the last file separator, while `dirname` returns everything up to, but not including, the last file separator.

```
> basename(toR)

[1] "R"

> dirname(toR)

[1] "/Users/robert/bin"
```

Exercise 4.6
Using the function `strsplit`, *write your own vectorized versions of both the* `basename` *function and the* `dirname` *functions. Include a* `sep` *argument that defaults to the value from* `.Platform`, *and use* `path.expand` *to handle input of the form* `~/foo`.

In the next example, we demonstrate how to create files, rename them, copy them and test for their existence. `file.append` uses the standard subscript recycling rules in trying to align its two arguments. Unfortunately, the order of the formal arguments that correspond to *from* and *to* is different for `file.append` than it is for all the other functions, so some caution is needed when using it.

```
> z = file.path(tempdir(), "foo")
> z
```

```
[1] "/tmp/RtmpHPmzRh/foo"

> file.create(tmp1)

[1] TRUE

> file.rename(tmp1, z)

[1] TRUE

> file.exists(tmp1, z)

[1] FALSE   TRUE

> file.copy(z, tmp1)

[1] TRUE

> file.exists(tmp1, z)

[1] TRUE TRUE

> file.symlink(z, tmp2)

[1] TRUE

> file.exists(tmp2)

[1] TRUE

> fiz = file.info(z)
> fitmp2 = file.info(tmp2)
> all.equal(fiz, fitmp2)

[1] "Attributes: < Component 2: 1 string mismatch >"
```

file.rename renames the file specified by its first argument with the name given as its second argument. Symbolic links can be created using the file.symlink function.

Lastly, one can create and manipulate directories themselves. The function dir.create will create a directory, which at that point can be manipulated like any other file. Note, however, that if the directory is not empty, file.remove will not remove it and will return FALSE. To remove directories that contain files, one must use the unlink function. unlink also works on plain files, but file.remove is probably more intuitive and slightly less dangerous. Note that incautious use of unlink can irretrievably remove important files.

In the example below, we demonstrate the use of some of these functions. We do most of the reading and writing to R temporary directory.

```
> newDir = file.path(tempdir(), "newDir")
> newDir

[1] "/tmp/RtmpHPmzRh/newDir"

> newFile = file.path(newDir, "blah")
> newFile

[1] "/tmp/RtmpHPmzRh/newDir/blah"

> dir.create(newDir)
> file.create(newFile)

[1] TRUE

> file.exists(newDir, newFile)

[1] TRUE TRUE

> unlink(newDir, recursive = TRUE)
```

Setting the `recursive` argument to `unlink` to `TRUE` is needed to remove non-empty directories. If this argument has its default value, `FALSE`, then the command would fail to remove a non-empty directory the same as `file.remove` does. Unix users will recognize this as the equivalent of typing `rm -r` from the command line, so **be careful!** You can remove files and directories that you did not intend to and they generally cannot easily be retrieved or restored.

The function `file.choose` prompts the user, interactively, to select a file, and then returns that file's name as a character vector. On Windows, users are presented with a standard file selection dialogue; on Unix-like operating systems, they are expected to type the name at the command line.

4.2.3 Working with R's binary format

R objects can be saved in a standard binary format, which is related to XDR (Eisler, 2006), and is platform independent. An arbitrary number of R objects can be saved into a single file using the `save` command. They can be reloaded into R using the `load` command. These files can be copied to any other computer and loaded into R without any translation. When an archive has been loaded, the return value of `load` is the name of all objects that were loaded.

Both `save` and `load` allow the caller to specify a specific environment in which to find the bindings or in which to store the restored bindings.

4.3   Connections

As indicated above, all data input and output can be performed via connections. Connections are basically an extension of the notion of file and provide a richer set of tools for reading and writing data. Connections provide an abstraction of an input data source. Using connections allows a function to work in essentially the same way for data obtained from a local file, an R character vector, or a file on the Internet. Connections are implemented using the S3 class system and the base class is *connection*, which different types of connections extend. There are both `summary` and `print` methods for connections.

The most commonly used connection is a file, which can be opened for reading or writing data. The set of possible values that can be specified for the `open` argument is detailed in the manual page. Other types of connections are the FIFO, pipe and socket. These are all described in some detail below. Connections can be used to read from zipped files, using one of `gzfile`, `bzfile` or `unz`, depending on what tool was used to compress the file. These connections can be supplied to `readLines` or `read.delim`, which then simply read directly from the compressed files.

Of some general interest is the function `showConnections` that will show all connections and their status. With the default settings, only user-created open connections are displayed. This can be helpful in ensuring that a connection is open and ready or for finding connections that have been opened and forgotten.

```
> showConnections(all = TRUE)
  description class      mode text     isopen   can read
0 "stdin"     "terminal" "r"  "text"   "opened" "yes"
1 "stdout"    "terminal" "w"  "text"   "opened" "no"
2 "stderr"    "terminal" "w"  "text"   "opened" "no"
3 "RIO.tex"   "file"     "w+" "text"   "opened" "yes"
4 ""          "file"     "w+" "text"   "opened" "yes"
  can write
0 "no"
1 "yes"
2 "yes"
3 "yes"
4 "yes"
```

Some connections support the notion of pushing character strings *back* onto the connection. One might presume that the function `pushBack` can only push back things that have been read; this is similar to the notion of rewinding a file, but this is not true. You can push back any character vector onto a connection that supports pushing back.

Not all operating systems support all connections. In order to determine whether your system has support for sockets, pipes or URLs the `capabilities` function can be used.

```
> capabilities()

    jpeg      png    tcltk      X11     aqua http/ftp  sockets
    TRUE     TRUE     TRUE     TRUE     TRUE     TRUE     TRUE
  libxml     fifo   cledit    iconv      NLS  profmem    cairo
    TRUE     TRUE    FALSE     TRUE     TRUE     TRUE    FALSE
```

4.3.1 Text connections

A text connection is essentially a device for reading from, or writing to, an R character vector. The code below is taken from the manual page for `textConnection` and it demonstrates some of the basic operations that can be carried out on a `textConnection` that is being used for input. The connection can be used as input for any of the input functions, such as `readLines` and `scan`, but it also supports pushing data onto the connection.

```
> zz = textConnection(LETTERS)
> readLines(zz, 2)

[1] "A" "B"

> showConnections(all = TRUE)

  description class              mode text   isopen   can read
0 "stdin"     "terminal"         "r"  "text" "opened" "yes"
1 "stdout"    "terminal"         "w"  "text" "opened" "no"
2 "stderr"    "terminal"         "w"  "text" "opened" "no"
3 "RIO.tex"   "file"             "w+" "text" "opened" "yes"
4 ""          "file"             "w+" "text" "opened" "yes"
5 "LETTERS"   "textConnection"   "r"  "text" "opened" "yes"
  can write
0 "no"
1 "yes"
2 "yes"
```

```
3 "yes"
4 "yes"
5 "no"

> scan(zz, "", 4)

[1] "C" "D" "E" "F"

> pushBack(c("aa", "bb"), zz)
> scan(zz, "", 4)

[1] "aa" "bb" "G"  "H"

> close(zz)
```

One can also write to a *textConnection*, and the effect is to create a character vector with the specified name; but you must be sure to use `open="w"` so that it is open for writing. You almost surely want to set `local=TRUE`; otherwise, the text connection is created in the top-level workspace. Since R's input and output can be redirected to a connection, this allows users to capture function output and store it in a computable form.

In the code below, we create a text connection that can be written to, then carry out some computations and use `sink` to divert the output of the command to the text connection. Since we did not set `local=TRUE`, creating the text connection creates a global variable named `foo`. We did set `split=TRUE` so that the output of the commands would be shown in the terminal and collected into the text connection. Other text can be written directly to the text connection using `cat` or other similar functions.

```
> savedOut = textConnection("foo", "w")
> sink(savedOut, split = TRUE)
> print(1:10)

 [1]  1  2  3  4  5  6  7  8  9 10

> cat("That was my first command \n")

That was my first command

> letters[1:4]

[1] "a" "b" "c" "d"

> sink()
> close(savedOut)
> cat(foo, sep = "\n")
```

Another alternative for capturing the output of commands is the suggestively named `capture.output`. Unlike using `sink`, the commands for which the output is wanted are passed explicitly to `capture.output`.

4.3.2 Interprocess communications

Being able to pass data from one process to another can lead to substantial benefits and should often be considered as an alternative to reimplementation or other, more drastic solutions. One of the more popular methods of sharing data between processes has been the use of intermediate files; one process writes a file and the other reads it. However, if the mechanics are left to the programmer, this procedure is fraught with danger and often fails in rather peculiar ways. Fortunately, there are a wide number of programmatic solutions that allow software to handle most of the organizational details, thereby freeing the programmer to concentrate on the conceptual details.

Some of the different connections and mechanisms for interprocess communication (IPC) have implementations as R connections, and we discuss those here. We also make some more general comments, and it is likely the future versions of R will include more refined IPC tools. A very sophisticated and detailed discussion of many of the concepts mentioned here is given in Stevens and Rago (2005), particularly Chapters 15 and 17 of that reference.

4.3.2.1 Socket connections

You can use the `capabilities` function to determine whether sockets and hence `socketConnections` are supported by your version of R. If they are, then the discussion in this section will be relevant. If they are not supported, then you will not be able to use them.

Sockets are a mechanism that can be used to support interprocess communications. Each of the two processes establishes a connection to a socket, which is merely one end of the intercommunication process. One process is typically the server and the other the client.

In the example below, we demonstrate how to establish a socket connection between two running R processes. For simplicity we presume that they are both running on the same computer, but that is not necessary; and in the general case, the processes can be on different computers. Furthermore, there is no requirement that both ends be R processes.

The default for socket connections is to be in non-blocking mode. That means that they will return as soon as possible. On input, they return with the available input, possibly nothing; and on output, they return regardless of whether the write succeeded.

The first R process sets up a socket connection on a named port in server mode. The port number is not important but you need to select one that is high enough not to conflict with a port that is in use.

```
serverCon = socketConnection(port = 6543, server=TRUE)
```

```
writeLines(LETTERS, serverCon)
close(serverCon)
```

Then, the second R process opens a connection to the same port but, this
time, in client mode. Since the client mode is not blocking, we must poll
until we have a complete input. The call to `Sys.sleep` ensures that some time
elapses between calls to `readLines` and allows other processes to be run.

```
clientCon = socketConnection(port = 6534)
readLines(clientCon)
while(isIncomplete(clientCon)) {
    Sys.sleep(1)
    readLines(clientCon)}
close(clientCon)
```

Unfortunately, connections are not exposed at the C level so there is no
opportunity for accessing them directly at that level.

4.3.2.2 Pipes

A pipe is a shell command where the standard input can be written from
R and the standard output can be read from R. A call to `pipe` creates a
connection that can be opened by writing to it, or by reading from it. The
pipe can be used as a connection for any function that reads and writes from
connections. In the example below, the system function `cal` is used to get a
calendar.

```
> p1 = pipe("cal 1 2006")
> p1

 description         class           mode           text
"cal 1 2006"        "pipe"           "r"          "text"
      opened     can read      can write
    "closed"         "yes"          "yes"

> readLines(p1)

[1] "    January 2006"      " S  M Tu  W Th  F  S"
[3] " 1  2  3  4  5  6  7"  " 8  9 10 11 12 13 14"
[5] "15 16 17 18 19 20 21"  "22 23 24 25 26 27 28"
[7] "29 30 31"              ""
```

It is reasonably simple to extend this to provide a function that returns
the calendar for either the current month, or any other month or year. The
function is provided in **RBioinf**, and the code is shown below.

```
> library("RBioinf")
> Rcal

function (month, year)
{
    pD = function(x) pipe(paste("date \"+%", x, "\"", sep = ""))
    if (missing(month))
        month = readLines(pD("m"))
    if (missing(year))
        year = readLines(pD("Y"))
    cat(readLines(pipe(paste("cal ", month, year))), sep = "\n")
}
<environment: namespace:RBioinf>
```

An alternative to the use of `pipe` is available using the `intern` argument for `system`. Notice that the following calls are equivalent. But `pipe` is more general, and `system` could easily be written in terms of `pipe`. Further, there is no real reason why a pipe cannot be bidirectional; Stevens and Rago (2005) refer to these as STREAMS-based pipes, which are opened for both reading and writing, but only unidirectional pipes have been implemented in R. Basically this means that to capture the output of any pipe opened for writing, you will need to redirect the output to file, or perhaps a socket or FIFO, and then read from that using a separate connection. On OS X, users can read and write from the system clipboard using `pipe("pbpaste")` and `pipe("pbcopy", "w")`, respectively.

```
> ww = system("ls -1", intern = T)
> xx = readLines(pipe("ls -1"))
> all.equal(ww, xx)

[1] TRUE
```

Another advantage to pipes over calls to `system` is that one can pass values to the system call via `pipe` after the subprocess has been started. With calls to `system`, the entire command must be assembled and sent at one time.

Exercise 4.7
Rewrite `Rcal` *to use* `system`.

Exercise 4.8
The following code establishes a `pipe` *to the system command* `wc`, *which counts words, characters and lines. What happens to the output? How would you modify the commands to retrieve the output of* `wc`?

```
WC = pipe("wc", open="w")
writeLines(letters,WC)
```

4.3.2.3   FIFOs

A FIFO is a special kind of file that stores data that are written to it. FIFO is an acronym for first-in, first-out, and FIFOs are also often referred to as named pipes. The next program to read the FIFO extracts the first record that was written, as the name suggests. Once a record has been read, it is automatically removed. Thus, the FIFO only retains data that have been written, but not yet read. FIFOs can thus be used for interprocess communication (as can `socketConnections`); but since FIFOs are named files, the communication channel is via the file system. Not all platforms support FIFOs. Most Unix-based versions and OS X do support `fifo`.

Pipes can only be used to communicate between processes that share a common ancestor that created the pipe. When unrelated processes want to communicate, they must make use of some other mechanism, and often the appropriate tool is a FIFO. Stevens and Rago (2005) give two different uses for FIFOs: first as a way for shell commands to pass data without creating intermediate temporary files and second as a rendezvous point for client-server applications to pass data between clients and servers.

4.3.3   Seek

Some connections support direct interactions with the location of the current read and write positions. If the connection supports these interactions, `isSeekable` will return `TRUE` and `seek` can be used to find the current position and to alter it. In the code chunk below, we create a file, write to it, and then manipulate the reading position using `seek`. Notice that the part of the file read is repeated. The connection is closed and the file is `unlinked` at the end of the code chunk.

```
> fName = file.path(tempdir(), "test1")
> file.create(fName)

[1] TRUE

> sFile = file(fName, open = "r+w")
> cat(1:10, sep = "\n", file = sFile)
> seek(sFile)

[1] 21

> readLines(sFile, 3)

[1] "1" "2" "3"
```

```
> seek(sFile, 2)

[1] 6

> readLines(sFile)

[1] "2"  "3"  "4"  "5"  "6"  "7"  "8"  "9"  "10"

> close(sFile)
> unlink(fName)
```

Thus, using seek, one can treat a large file as random access memory. However, the cost can be quite high as reading and writing tends to be a bit slow. Other alternatives are to read the data in and use internal tools or database tools such as those described in Chapter 8.

4.4   File input and output

The most appropriate tool for reading and writing from files will generally depend on the contents of the file, and the purpose to which those contents will be put. The most general low-level reading function is scan. Perhaps two of the most general commands for file input/output are readLines and writeLines. As their names suggest, the former is used to read input and the latter to write output. Both functions take a parameter con, which will take either the name of a file or a connection. The default for this parameter is to read/write from stdin and stdout, respectively. From here on, however, they differ.

readLines has the following formal arguments:

n the (maximal) number of lines to read. Negative values indicate reading to the end of the connection. Default is -1.

ok a logical value indicating whether it is OK to reach the end of the connection before n > 0 lines are read. If not, an error will be generated. The default of this is TRUE.

warn a logical value indicating whether or not to warn the user if a text file is missing a final end-of-line character.

encoding the encoding that is assumed for the input.

writeLines has the following formal arguments:

text a character vector.

sep a string to be written to the connection after each line of text. Default is
the new line character, `"\n"`.

```
> a = readLines(con = system.file("CONTENTS", package = "base"),
+       n = 2)
> a

[1] "Entry: Arithmetic"
[2] "Aliases: + - * ** / ^ %% %/% Arithmetic"

> writeLines(a)

Entry: Arithmetic
Aliases: + - * ** / ^ %% %/% Arithmetic
```

A rather frequent question on the R mailing list is how to create files and
write to those files within a loop. For example, suppose that there is some
interest in carrying out a permutation test and saving the permutations in
separate files. In the code below, we show how to do this for a small example
with 10 permutations. The files are written into the temporary directory in
this example so that they will be removed when R exits. You should choose
some other location to write to, but that will depend on your local file system.

```
> mydir = tempdir()
> for (i in 1:10) {
+       fname = paste("perm", i, sep = "")
+       prm = sample(1:10, replace = FALSE)
+       write(prm, file = file.path(mydir, fname))
+ }
```

Exercise 4.9
*Select a location on your local file system and use the code above to write files
in that location. How would you modify the code to write a comma-separated
set of numbers? What seed was used to generate the permutations? Can you
set a seed so you always get the same permutations?*

4.4.1   Reading rectangular data

In many cases the data to be read in are in the form of a rectangular,
or nearly rectangular, array. For those cases, there are specialized functions (`read.table`, `read.delim` and `read.csv`) with variants (`read.csv2` and
`read.delim2`) that are tailored to European norms for representing numbers.

These functions will take either the name of a file or a connection and attempt to read data from that. There are three primary ways in which they differ: what is considered to be a separator of the data items, the character used to delimit quoted strings, and what character is used for the decimal indicator. The most general of these is `read.table` and, in fact, the others are merely wrappers to `read.table` with appropriate values set for the arguments. However, comma-separated values (`.csv`) occur often enough that it is worthwhile to have the convenience function.

Among the more important arguments to `read.table` are:

`as.is` by default, character variables are turned into factors; setting `as.is` to `TRUE`, they are left as character values. The transforming of strings into factors can be controlled using the `option` `stringsAsFactors`.

`na.strings` a vector of strings that are to be interpreted as missing values and hence any corresponding entries will be converted to `NA` during processing.

`fill` if set to `TRUE` and some rows have unequal lengths, shorter rows are padded.

`comment` a single character indicating the *comment character.* for any line of the input file, all characters after the comment character are skipped.

`sep` the record separator.

`header` a logical value indicating whether or not the first line of the file contains the variable names.

When the data do not appear to be read in correctly, the three most common causes are: the quote character is used in the file for something other than a quotation, and hence the symbols are not matched (for biological data 3' and 5' are often culprits); the comment character appears in the file, not as a comment; or there are some characters in the file that have an unusual encoding and have caused confusion.

The default behavior of these different routines is to turn character variables (columns) into factors. If this is not desired, and very often it is not, then either the `as.is` argument should be used or the more general `colClasses` should be used. `colClasses` can be used to specify classes for all columns. If a column has a `NULL` value for `colClasses`, then that column is skipped and not read into R.

4.4.2 Writing data

Since R's roots are firmly in statistical applications where data have a rectangular form, typically with rows corresponding to cases and columns to variables, there are specialized tools for writing rectangular data. Some of these are aimed at producing a table that is suitable for being imported into a spreadsheet application such as Gnumeric or Microsoft's Excel.

The function `write` can be used to write fairly arbitrary objects. While it has a number of arguments that are useful for writing out matrices, it does not deal with data frames. For writing out data frames, there are three separate functions: the very general `write.table` and two specialized interfaces to the same functionality, `write.csv` and `write.csv2`.

Another way to write R objects to a file is with the function `cat`. The transformation of R objects to character representations suitable for printing is different from those carried out by either `write` or `print`. By default, `cat` writes to the standard output connection, but the `file` argument can be any connection, the name of a file or the special form `"|cmd"`, in which case the output of `cat` is sent to the system function named. In the code chunk below, we use this feature to send the output of `cat` to the `cal` function.

```
> cat("10 2005", file = "|cal")
```

Other functions that are of interest include `writeBin` and `readBin` for reading and writing binary data, as well as `writeChar` and `readChar` for reading and writing character data. Readers are referred to the relevant manual pages and the R Data Import/Export Manual for more details on these functions.

4.4.3   Debian Control Format (DCF)

Debian Control Format (DCF) is used for some of the package-specific files in R (see Chapter 7); in particular, the DESCRIPTION file in all R packages and the CONTENTS file for installed packages. The functions `read.dcf` and `write.dcf`, are available in R to read and write files in this format. For a description of DCF, see `help("read.dcf")`.

```
> x = read.dcf(file = system.file("CONTENTS", package = "base"),
+      fields = c("Entry", "Description"))
> head(x, n = 3)

      Entry
[1,] "Arithmetic"
[2,] "AsIs"
[3,] "Bessel"
      Description
[1,] "Arithmetic Operators"
[2,] "Inhibit Interpretation/Conversion of Objects"
[3,] "Bessel Functions"
```

```
> write.dcf(x[1:3, ], file = "")
```

Entry: Arithmetic
Description: Arithmetic Operators

Entry: AsIs
Description: Inhibit Interpretation/Conversion of
      Objects

Entry: Bessel
Description: Bessel Functions

read.dcf returns a matrix, while write.dcf takes a matrix and transforms it into DCF formatted output; the empty string "" as the file parameter tells the system to output to the console instead of specifying a particular file.

4.4.4 FASTA Format

Biological sequence data are available in a very wide range of formats. The FASTA format is probably the most widely used, but there are many others. A FASTA file consists of one or more biological sequences. Each sequence is preceded by a single line, beginning with a >, which provides a name and/or a unique identifier for the sequence and often other information. The description line can be followed by one or more comment lines, which are distinguished by a semicolon at the beginning of the line. After the header line and comments, the sequence is represented by one or more lines. Sequences may correspond to protein sequences or DNA sequences and should make use of the IUPAC codes; these can be found in many places, including http://en.wikipedia.org/wiki/Fasta_format. All lines should be shorter than 80 characters. Functions for reading and writing in the FASTA format are provided in the **Biostrings** package as readFASTA and writeFASTA, respectively.

Exercise 4.10
*Modify the function readFASTA in the **Biostrings** package, or any other FASTA reading function, to (1) transform the data to uppercase, (2) check that only IUPAC symbols are contained in the sequence data, and (3) check the line lengths to see if they are shorter than 80 characters.*

Exercise 4.11
There is a file in the Biostrings package, in a folder named extdata, named exFASTA.mfa. Using the system.file and readLines functions, process this file to answer the following questions. How many records are in the file? How long, in number of characters, are the different records? Can you tell if they are DNA or protein sequences that have been encoded?

Compare your approach with that in the readFASTA *function. What are the differences? Run a timing comparison to see which is faster (you might want to refer to Section 9.5 for details on how to do that).*

4.5   Source and sink: capturing R output

While the standard interactions with R are primarily carried out by users typing commands to the command line and subsequently viewing the outputs that those commands generate, there are many situations where more programmatic interactions are important. Often, users will want to either supply input to R in some other way, or they may want to capture the output of a command into a file or variable so that it can be programmatically manipulated or simply for future reference.

The main interface for input is source, which reads, parses and evaluates R commands. The input can be a file or a connection. Since the input is parsed, there is generally code rearrangement and in particular, by default, comments are dropped. The argument keep.source can be used to override this behavior, and there is also a global option, of the same name, that can be used to set behavior for the entire session. When source is run, it first reads, then parses, and finally evaluates, so no command will be evaluated if there is a syntax error in the file.

Users can also carry out the three steps themselves, if they choose. They can first use scan to read in the commands as text; then use parse to parse but not evaluate those commands; and finally use eval to evaluate the set of parsed expressions. Such a strategy allows for much more fine-grained control over the process, although it is seldom ever needed.

In other cases it will be quite helpful to capture the output of different commands issued to R. One example is the function printWithNumbers discussed in Chapter 9, which provides appropriate line numbers for R functions so that the at argument for trace can be more easily used. To implement this function, we used capture.output, which can be used to capture, as text, the output of a set of provided R expressions.

Alternatively, R's standard output can be diverted using sink. A call to sink will divert all standard output, but not error or warning messages, nor one presumes other conditions (e.g. Section 2.11). To divert error and warning messages, set the argument type to "messages". This capability should be used with caution, however, since it will not be easy to determine when errors are being signaled. Note that the requirements for the file argument are different when the messages are being diverted, than when output is being diverted. To turn off the redirection, simply call sink as second time with a NULL argument. The redirections established by sink form a stack, with new redirections added to the top, and calls to `sink` with `NULL` arguments popping the top redirection off of the stack. The function `sink.number` can be used to determine how many redirections are in use. It is also possible to both capture the output, via a redirection, and to continue to have it displayed on the screen. This is achieved by setting the `split` argument in the call to `sink`.

4.6 Tools for accessing files on the Internet

R functions that can be used to obtain files from the Internet include `download.file` and `url`, which opens a connection and allows for reading from that connection. The function `url.show` renders the remote file in the console.

There are R functions for encoding URLs, `URLencode` and `URLdecode`, that can be used to encode and decode URL names. URLs have a set of reserved characters, and not all characters are valid. An invalid character needs to be preceeded by a % sign if it is contained in a URL.

The **RCurl** package provides an extensive interface to `libcURL`. The `libcURL` library supports transferring files via a wide variety of protocols including FTP, FTPS, HTTP, HTTPS, GOPHER, TELNET, DICT, FILE and LDAP. `libcURL` also supports HTTPS certificates, HTTP POST, HTTP PUT, FTP uploading, Kerberos, HTTP form-based upload, proxies, cookies, user+password authentication, file transfer resume, and http proxy tunneling.

This package supports a very wide range of interactions with web resources and in particular is very helpful in posting and retrieving data from forms. Many bioinformatic databases and tools provide forms-based interfaces. These are often used interactively, by basically pointing a browser to the appropriate page and filling in values. However, one can post and retrieve answers programmatically. Alternatively, many provide Biomart interfaces, and the tools described in Section 8.6.3 can be used to obtain the data.

RCurl can eliminate manual work ("screen-scraping") with web pages to obtain data that have not been made available through standard web services. For example, when data can be obtained interactively using text input, radio button settings, and check-box selections, code resembling the following can be used to obtain that same data programmatically:

```
> postForm("http://www.speakeasy.net/main.php",
+           "some_text" = "Duncan", "choice" = "Ho",
+           "radbut" = "eep", "box" = "box1, box2" )
```

The resulting data must be parsed, but the `htmlTreeParse` function can be very helpful for this. More details on XML and HTML parsing are given in

Section 8.5.

The next example and its solution are based on a discussion from the R help mailing list. The Worldwide Protein Data Bank (wwPDB) is an online source for PDB data. The mission of the wwPDB is to maintain a single Protein Data Bank (PDB) Archive of macromolecular structural data that is freely and publicly available to the global community. The web site is at `http://www.wwpdb.org/`, and data can be downloaded from that site. However, there are an enormous number of files, and one might want to be somewhat selective in downloading. The code below shows how to obtain all the file names. Individual files can then be obtained using `wget` or other similar functions. The calls to `strsplit` and `gsub` split the string on the new line character and remove any `\r` (carriage return) characters that are present. We could have done that in one step by using a regular expresssion (Section 5.3) but then `strsplit` becomes painfully slow.

```
> library("RCurl")
> url = "ftp://ftp.wwpdb.org/pub/pdb/data/structures/all/pdb/"
> fileNames = getURL(url,
+       .opts = list(customrequest = "NLST *.gz") )
> fileNames = strsplit(fileNames, "\n", fixed=TRUE)[[1]]
> fileNames = gsub("\r", "", fileNames)
> length(fileNames)

[1] 51261
```

The file names are informative, as they encode PDB identifiers, and given a map to these, say from some genes of interest, perhaps using **biomaRt**, Section 8.6.3, one can download individual files of interest. In the code below, we download the first file in the list using `download.file`.

```
> fileNames[1]

[1] "pdb100d.ent.gz"

> download.file(paste("ftp://ftp.wwpdb.org/pub/pdb/data/",
+       "structures/all/pdb/pdb100d.ent.gz", sep = ""),
+       destfile = "pdb100d.ent.gz")
```

Chapter 5

Working with Character Data

5.1 Introduction

Working with character data is fundamental to many tasks in computational biology, but is not that common of a problem in statistical applications. The tools that are available in R are more oriented to the processing of data into a form that is suitable for statistical analysis, or to format outputs for publication. There is an increased awareness, and corresponding capabilities, for dealing with different languages and file encodings, but we will not do more than briefly touch on this subject. In this chapter we review the builtin capabilities in R, but then turn our attention to some problems that are more fundamental to biological applications.

In biological applications there are a number of different alphabets that are relevant; perhaps the best known of them is the four letter alphabet that describes DNA, but there are others. The basic problems are exact matching of one or more query sequences in a target sequence, inexact matching of sequences, and the alignment of two or more sequences. There is also substantial interest in text mining applications, but we will not cover that subject here. Our primary focus will be on the methodology provided in the **Biostrings** package, but there are a number of other packages that have a biological focus, including **seqinR**, **annotate**, **matchprobes**, **GeneR** and **aaMI**.

String matching problems exist in many different contexts, and have received a great deal of attention in the literature. Cormen et al. (1990) provide a nice introduction to the methods while Gusfield (1997) gives a much more in depth discussion with many biological applications. One can either search for exact matches of one string in another, or for inexact matches. Inexact matching is more difficult and often more computationally expensive.

The chapter is divided into three main sections. First we describe the builtin functions for string handling and manipulation, plus some generic string handling issues. Next we discuss regular expressions and tools such as `grep` and `agrep`, and finally we present more detail on the biological problems and present a number of concrete examples.

5.2   Builtin capabilities

Character vectors are one of the basic vector types in R (see Chapter 2 for more details). A character vector consists of zero or more character strings. Only the vector can be easily manipulated at the R level and most functions are vectorized. The string `"howdy"` will be stored as a length one character vector with the first element having five characters. In the code below, we construct a character vector of length three. Then we use `nchar` to ask how many characters there are in each of the strings. The function `nchar` returns the length of the elements of its argument. There are three different ways to measure length: `bytes`, `chars` and `width`. These are generally the same, at least in locales with single-byte characters.

```
> mychar = c("as", "soon", "as possible")
> mychar

[1] "as"            "soon"           "as possible"

> nchar(mychar)

[1]  2  4 11
```

Like other basic types, a character vector can be of length zero; and in the code below we demonstrate the difference between a character vector of length zero and a character string of length zero. The variable x represents a zero length character vector, while y represents a length one character vector, whose single element is the empty string.

```
> x = character(0)
> length(x)

[1] 0

> nchar(x)

integer(0)

> y = ""
> length(y)

[1] 1

> nchar(y)

[1] 0
```

To access substrings of a character vector, use either `substr` or `substring`. These two functions are very similar but handle recycling of arguments differently. The first three arguments are the character vector, a set of starting indices and a vector of ending indices. For `substr`, the length of the returned value is always the length of its first argument (`x`). For `substring`, it is the length of the longest of these three supplied arguments; the other arguments are recycled to the appropriate length.

```
> substr(x, 2, 4)

character(0)

> substr(x, 2, rep(4, 5))

character(0)

> substring(x, 2, rep(4, 5))

character(0)
```

A biological application of `substring` is to build a function to translate DNA into the corresponding amino acid sequence. We can use `substring` to split an input DNA sequence into triples, which are then used to index into the `GENETIC_CODE` variable, and finally we paste the amino acid sequences together. The `GENETIC_CODE` variable presumes that the sequence given is the sense strand.

```
> rD = randDNA(102)
> rDtriples = substring(rD, seq(1, 102, by = 3),
+      seq(3, 102, 3))
> paste(GENETIC_CODE[rDtriples])

 [1] "V" "R" "N" "Y" "P" "S" "K" "A" "L" "C" "*" "Q" "V"
[14] "A" "C" "L" "Q" "*" "S" "N" "M" "D" "D" "L" "Q" "Q"
[27] "L" "S" "N" "L" "V" "C" "L" "H"
```

Exercise 5.1
Using the code above, create a simple function that maps from DNA to the amino acid sequence.

It is also possible to modify a string and the replacement versions of `substr` and `substring` are used for this purpose. In the example below, we demonstrate some differences between the two functions. These functions are evaluated for their side effects, which are changes to the character strings contained

in their first argument. There are no default values for either the starting position or the ending position in `substr`. For `substring` there is a default value
for the stopping parameter.

```
> substring(x, 2, 4) = "abc"
> x

character(0)

> x = c("howdy", "dudey friend")
> substr(x, 2, 4) = "def"
> x

[1] "hdefy"           "ddefy friend"

> substring(x, 2) <- c("..", "+++")
```

Exercise 5.2
What happens if the stop, or last, argument to `substr` *or* `substring` *is larger
than the number of characters? Is it different for the replacement version? In
the replacement version, what happens if the length of the string to assign is
longer than the character vector.*

Character strings can be appended using `paste`. `paste` takes any number of
arguments and coerces them to be character vectors first. The usual recycling
rules for function arguments apply, in that all arguments are treated as vectors,
and if necessary those that are shorter than the longest supplied argument are
replicated until all arguments are of length equal to the longest input vector.
The return value is a character vector of length equal to that of the longest
input, with all inputs concatenated.

In the code below, we `paste` three vectors; the first is of length three, the
second of length one, and the third of length two. They are replicated different
times to yield a vector of length three (the length of the longest input). In
the second example, we demonstrate the use of the `sep` argument.

```
> paste(1:3, "+", 4:5)

[1] "1 + 4" "2 + 5" "3 + 4"

> paste(1:3, 1:3, 4:6, sep = "+")

[1] "1+1+4" "2+2+5" "3+3+6"
```

In some cases, the desire is to reduce a character vector with multiple character strings to one with a single character string, and the `collapse` argument can be used to reduce, or collapse, the input vector.

```
> paste(1:4, collapse = "=")

[1] "1=2=3=4"
```

The reverse operation, that of splitting a long string into substrings, is performed using the `strsplit` function. `strsplit` takes any character string or a regular expression as the splitting criterion and returns a list, each element of which contains the splits for the corresponding element of the input. If the input string is long, be sure to either use Perl regular expressions or set `fixed=TRUE`, as the standard regular expression code is painfully slow. To split a string into single characters, use the empty string or `character(0)`. While the help page recommends the use of either `character(0)` or `NULL`, these can be problematic if the second argument to `strsplit` is of length more than one. Compare the two outputs in the example below.

```
> strsplit(c("ab", "cde", "XYZ"), c("Y", ""))

[[1]]
[1] "ab"

[[2]]
[1] "c" "d" "e"

[[3]]
[1] "X" "Z"

> strsplit(c("ab", "cde", "XYZ"), c("Y", NULL))

[[1]]
[1] "ab"

[[2]]
[1] "cde"

[[3]]
[1] "X" "Z"
```

It is sometimes important to output text strings so that they look nice on the screen or in a document. There are a number of functions that are available,

and we have produced yet another one that is designed to interact with the
Sweave system. Two builtin functions are `strtrim`, which trims strings to a
fixed width, and `strwrap`, which introduces line breaks into a text string.

To trim strings to fit into a particular width, say for text display, use
`strtrim`. The arguments to `strtrim` are the character vector and a vector
of widths. The widths are interpreted as the desired width in a monospaced
font. To wrap text use `strwrap`, which honors a number of arguments including
the width, indentation, and a user-supplied prefix.

```
> x <- paste(readLines(file.path(R.home(), "COPYING")),
+     collapse = "\n")
> strwrap(x, 30, prefix = "myidea: ")[1:10]

 [1] "myidea: GNU GENERAL PUBLIC"
 [2] "myidea: LICENSE Version 2,"
 [3] "myidea: June 1991"
 [4] "myidea: "
 [5] "myidea: Copyright (C) 1989,"
 [6] "myidea: 1991 Free Software"
 [7] "myidea: Foundation, Inc.  51"
 [8] "myidea: Franklin St, Fifth"
 [9] "myidea: Floor, Boston, MA"
[10] "myidea: 02110-1301 USA"

> writeLines(strwrap(x, 30, prefix = "myidea: ")[1:5])

myidea: GNU GENERAL PUBLIC
myidea: LICENSE Version 2,
myidea: June 1991
myidea:
myidea: Copyright (C) 1989,
```

When using `Sweave` to author documents, such as this book, the author will
often need to ensure that no output text string is wider than the margins.
While one might anticipate `strwrap` would facilitate such requests, it does
not. We have written a separate simple function, `strbreak`, in the **Biobase**
package, to carry out this task.

Exercise 5.3
*Compare the function `strbreak` with `strwrap` and `strtrim`. What are the
differences in terms of the output generated?*

5.2.1 Modifying text

Text can be transformed; calls to `toupper` and `tolower` change all characters in the supplied arguments to upper case and lower case, respectively. Non-alphabetic characters are ignored by these two functions. For general translation from one set of characters to another, use `chartr`. In the code chunk below we present a small function to translate from the DNA representation to the RNA representation. Basically, DNA is represented as a sequence of the letters A, C, T, G, while for RNA, U is substituted for T. We first transform the input to upper case, and then use `chrtr` to transform all instances of T into U. Notice that the function is vectorized, since we have only made use of functions that are themselves vectorized. We use the `randDNA` function to generate random DNA strings.

```
> dna2rna = function(inputStr) {
+       if (!is.character(inputStr))
+           stop("need character input")
+       is = toupper(inputStr)
+       chartr("T", "U", is)
+ }
> x = c(randDNA(15), randDNA(12))
> x

[1] "TCATCCATTCGTGGG" "GTTGGTCCATAG"

> dna2rna(x)

[1] "UCAUCCAUUCGUGGG" "GUUGGUCCAUAG"
```

Exercise 5.4
Write a function for translating from RNA to DNA. Test it and dna2rna *on a vector of inputs.*

The function `chartr` can translate from one set of values to another. Hence it is simple to write a function that computes the complementary sequence for either DNA or RNA.

```
> compSeq = function(x) chartr("ACTG", "TGAC",
+       x)
> compSeq(x)

[1] "AGTAGGTAAGCACCC" "CAACCAGGTATC"
```

Exercise 5.5
Write a function to test whether a sequence is a DNA sequence or an RNA sequence. Modify the function compSeq *above to use the test and perform the appropriate translation, depending on the type of input sequence.*

Users can also use sub and gsub to perform character conversions, and these functions are described more fully in Section 5.3. One limitiation of chartr is that it does strict exchange of characters, and for some problems you will want to either remove characters or replace a substring with a longer or shorter substring, which cannot be done with chartr but can be done with sub or gsub.

While complement sequences are of some interest in biological applications, reverse complementing is more common as it reflects the act of transcription. Tools for performing this manipulation on DNA and RNA sequences are provided in the matchprobes and Biostrings packages.

Exercise 5.6
Look at the manual page for strsplit *to get an idea of how to write a function that reverses the order of characters in the character strings of a character vector. Use this to write a reverseComplement function.*

5.2.2   Sorting and comparing

The basis for ordering of character strings is lexicographic order in the current locale, which can be determined by a call to Sys.getlocale. Comparisons are done one character at a time; if one string is shorter than the other and they match up to the length of the shorter string, the longer string will be sorted larger. The arithmetic operators <, >, ==, and != can all be applied to character vectors. And hence other functions such as max, min and order can also be used.

```
> set.seed(123)
> x = sample(letters[1:10], 5)
> x

[1] "c" "h" "d" "g" "f"

> sort(x)

[1] "c" "d" "f" "g" "h"

> x < "m"

[1] TRUE TRUE TRUE TRUE TRUE
```

5.2.3 Matching a set of alternatives

Searching or matching a set of input character strings in a reference list or table can be performed using one of `match`, `pmatch` or `charmatch`. Each of these has different capabilities, but all work in a more or less similar manner. The first argument is the set of strings that matches are desired for; the second is the table in which to search. The returned value from these functions is a vector of the same length as the first argument that contains the index of the matching value in the second argument, or the value of the `nomatch` parameter if no match is found. The function `%in%` is similar to `match` but returns a vector of logical values, of the same length as its left operand indicating which elements were found in the right operand. The first argument (left operand in the case of `%in%`) is converted to a character vector (using `as.character`) prior to evaluation.

```
> exT = c("Intron", "Exon", "Example", "Chromosome")
> match("Exon", exT)

[1] 2

> "Example" %in% exT

[1] TRUE
```

Both `pmatch` and `charmatch` perform partial matching. Partial matching is similar to that used for arguments to functions, where matching is done per character, left to right. For both functions, the elements of the first argument are compared to the values in the second argument. First, exact matches are determined. Then, any remaining arguments are tested to see if there is an unambiguous partial match and, if so, that match is used. By default, the elements of the `table` argument are used only once; for `pmatch`, this behavior can be changed by setting the `duplicates.ok` argument to `TRUE`. These functions do not accept regular expressions. For matching using regular expressions, see the discussion in Section 5.3.

The functions differ in how they deal with non-matches versus ambiguous partial matches, but otherwise are very similar. With `pmatch`, the empty string, `""` matches nothing, not even the empty string, while with `charmatch` it does match the empty string. `charmatch` reports ambiguous partial matches as 0 and non-matches as `NA`, while `pmatch` uses `NA` for both.

In the example below, the first partial match fails because two different values in `exT` begin with a capital E. The second call identifies the second element since enough characters were supplied to uniquely identify that value. The third example succeeds since there is only one value in `exT` that begins with a capital I, and the fourth example demonstrates the use of the very similar function `charmatch`..

```
> pmatch("E", exT)

[1] NA

> pmatch("Exo", exT)

[1] 2

> pmatch("I", exT)

[1] 1

> charmatch("I", exT)

[1] 1
```

Exercise 5.7
Test the claims made above about matching of the empty string; show that with pmatch *there is no match, while with* charmatch *there is.*

The behavior is a bit different if multiple elements of the input list match a single element of the table, versus when one element of the input list matches multiple elements in the table. In the first example below, even though more characters matched for the second string, it is not used as the match; thus all partial matches are equal, regardless of the quality of the partial match. Using either duplicates.ok=TRUE or charmatch will find all partial matches in the table.

```
> pmatch(c("I", "Int"), exT)

[1]  1 NA

> pmatch(c("I", "Int"), exT, duplicates.ok = TRUE)

[1] 1 1

> charmatch(c("I", "Int"), exT)

[1] 1 1
```

If there are multiple exact matches of an input string to the table, then pmatch returns the index of the first, while charmatch returns 0, indicating ambiguity.

```
> pmatch(c("ab"), c("ab", "ab"))

[1] 1

> charmatch(c("ab"), c("ab", "ab"))

[1] 0
```

5.2.4 Formatting text and numbers

Formatting text and numbers can be accomplished in a variety of different ways. Formatting character strings or numbers, including interpolation of values into character strings, can be accomplished using `paste` and `sprintf`. Formatting of numbers can be achieved using either `format` or `formatC`. Use the **xtable** package for formatting R objects into LaTeX or HTML tables. The function `sprintf` is an interface to the C routine `sprintf`, which supports all of the functionality of that routine, with R-style vectorization. The function `formatC` formats numbers using C style format specifications. But it does so on a per-number basis; for common formatting of a vector of numbers, you should use `format`. `format` is a generic function with a number of specialized methods for different types of inputs, including matrices, factors and dates.

5.2.5 Special characters and escaping

A string literal is a notation for representing sets of characters, or strings, within a computer language. In order to specify the extent of the string, a common solution is the use of delimiters. These are usually quotation marks and in R either single, ', or double, ", quotes can be used to delimit a string. The delimiters are not part of the string, so the problem of how to have a string with either a single or double quote in it arises. In one sense this is easy to solve, since strings delimited with double quotes can contain a single quote, and vice versa, but that does not entirely preclude the need for a mechanism for indicating that a character is to be treated specially. A fairly widely used solution is the use of an escape character. The meaning of the escape character is to convey the intention that the next character be treated specially. In R, the escape character is the backslash, \.

Both strings below are valid inputs in R, and they are two distinct literals representing the same string.

```
> 'I\'m a string'

[1] "I'm a string"
```

```
>   "I'm a string"

[1] "I'm a string"
```

The next problem that arises is how to have the escape character appear in a string. But we have essentially solved that problem too: simply escape the escape character.

```
> s = "I'm a backslash: \\"
> s

[1] "I'm a backslash: \\"
```

The printed value shows the escape character. That is because the `print` function shows the string literal that this variable is equal to, in the sense that it could be copied into your R session and be valid. To see the string literal, you can use `cat`. Notice that there are no quotes and that only one backslash appears in the output.

```
> cat(s)

I'm a backslash: \
```

You can print a string without additional quotes around it using the `noquote` function, but that is not the same as using `cat`; you will still see the R representation of the string. Notice in the example that there is a double backslash printed, unlike the output of `cat`.

```
> noquote(s)

[1] I'm a backslash: \\
```

Special characters represent non-printing characters, such as new lines and tabs. These control characters are single characters. You can check this using the function `nchar`. Octal and hexidecimal codes require an escape as well. More details are given in Section 10.3.1 of R Development Core Team (2007b).

```
> nchar(s)
```

```
[1] 18

> strsplit(s, NULL)[[1]]

 [1] "I"  "'"  "m"  " "  "a"  " "  "b"  "a"  "c"  "k"
[11] "s"  "l"  "a"  "s"  "h"  ":"  " "  "\\"

> nchar("\n")

[1] 1

> charToRaw("\n")

[1] 0a
```

The backslash was not escaped and so it is interpreted with its special
meaning in the third line, and R correctly reports that there is a single character. On the fourth line, we convert the character code into *raw bytes* and see
the ASCII representation for the new line character. All would be relatively
easy, except that the backslash character sometimes gets used for different
things; and on Windows, it turns out to be the file separator. Even that is
fine, although when creating pathnames in R, you must remember to escape
the backslashes, as is done in the example below. Of course, one should use
both `file.path` and `system.file` to construct file paths and then the correct
separator is used.

```
> fn = "c:\\My Documents\\foo.bar"
> fn

[1] "c:\\My Documents\\foo.bar"
```

Now, if there is a desire to change the backslashes to forward slashes, that
can be handled by a number of different R functions such as either `chartr` or
gsub.

```
> old = "\\"
> new = "/"
> chartr(old, new, fn)

[1] "c:/My Documents/foo.bar"
```

With `gsub`, the solution is slightly more problematic, since the string created
in R will be passed to another program that also requires escaping. In the first

call to `gsub` below, we must *double* each backslash so that the string, when passed to the Perl regular expression library (PCRE), has the backslashes escaped. In the second line, where we state that `fixed=TRUE`, only one escape is needed.

```
> gsub("\\\\", new, fn)

[1] "c:/My Documents/foo.bar"

> gsub("\\", new, fn, fixed = TRUE)

[1] "c:/My Documents/foo.bar"
```

5.2.6   Parsing and deparsing

Parsing is the act of translating a textual description of a set of commands into a representation that is suitable for computation. When you type a set of commands at the console, or read in function definitions from a file, the parser is invoked to translate the textual description into the internal representation. The inverse operation is called deparsing – which turns the internal representation into a text string. In the code below, we first parse a simple function call, and then show that the parsed value is indeed executable in R and then deparse it to get back a text representation. The parsed quantity is an *expression* and

```
> v1 = parse(text = "mean(1:10)")
> v1

expression(mean(1:10))

> eval(v1)

[1] 5.5

> deparse(v1)

[1] "expression(mean(1:10))"

> deparse(v1[[1]])

[1] "mean(1:10)"
```

Other functions that are commonly used for printing or displaying data are cat, print and show. In order to control the width of the output string, either strwrap or strtrim can be used.

5.2.7 Plotting with text

When creating a plot, one often wants to add text to the output device. Our treatment is quite cursory since there are other more comprehensive volumes (Murrell, 2005; Venables and Ripley, 2002) that deal with the topic of plotting data and working with the R graphics system.

We would like to draw attention to the notion of tool-tips. An implementation of them is in the imageMap function of the **geneplotter** package, which creates an HTML page and a MAP file that, when rendered in a browser, has user-supplied tool-tips embedded.

5.2.8 Locale and font encoding

String handling is affected by the locale and indeed what is a valid character, and hence what is a valid identifier in R is determined by the locale. Locale settings facilitate the use of R with different alphabets, monetary units and times. The locale can be queried and set using Sys.getlocale and Sys.setlocale.

```
> Sys.getlocale()


[1] "C"
```

These capabilities have been greatly expanded in recent versions of R, and many users in countries with multi-byte character sets, e.g., UTF-8 encodings, are able to work with those encodings. We will not cover these issues here. Users who want to explore native language support should examine the functions iconv and gettext. The former translates strings from one encoding into another while the latter describes the tools R uses to translate error and warning messages. Section 1.9 of R Development Core Team (2007c) should also be consulted.

5.3 Regular expressions

Regular expressions have become widely used in applied computing, spawning a POSIX standard as well as a number of books, including Friedl (2002) and Stubblebine (2007). Their uses include validation of input sequences, such as email addresses and genomic sequences, as well as a variety of search and optionally replace problems such as finding words or sentences with specific beginnings or endings. A regular expression is a pattern that describes a set of character strings.

In R, there are three different types of regular expressions that you can use: extended regular expressions, basic regular expressions and Perl-like regular expressions. The first two types of regular expressions are implemented using `glibc`, while the third is implemented using the Perl-compatible regular expressions (PCRE) library. We will present a view of the capabilities in R that is based on the description in the manual page for regular expressions which you can access via the command, `?regex`, that is itself based on the manual pages for GNU grep and the PCRE manual pages.

Among the functions that facilitate the use of regular expressions are `grep`, `sub`, `gsub regexp` and `gregexpr`. While `agrep` provides some similar capabilities, it does not use regular expressions, but rather depends on metrics between strings. The functions `strsplit`, `apropos` and `browseEnv` also allow the use of regular expressions. In the examples below, we mainly use `regexpr` and `gregexpr` since they show both the position and the length of the match, and that is pedagogically useful.

We do not have the space to cover all possible uses or examples of regular expressions and rather focus on those tasks that seem to recur often in handling biological strings. Readers should consult the R manual pages, any of the many books (Friedl, 2002; Stubblebine, 2007), or online resources dedicated to regular expressions for more details.

5.3.1   Regular expression basics

All letters and digits, as well as many other single characters, are regular expressions that match themselves. Some characters have special meaning and are referred to as meta-characters. Which characters are meta-characters depends on the type of regular expression. The following are meta-characters for extended regular expressions and for Perl regular expressions: `. \ | ( ) [ { ^ $ * + ?`. For basic regular expressions, the characters `? { | ( )`, and `+` lose their special meaning and will be matched like any other character. Any meta-character that is preceded by a backslash is said to be *quoted* and will match the character itself; that is, a quoted meta-character is not interpreted as a meta-character. Notice that in the discussion in Section 5.2.5, we referred to essentially the same idea as *escaping*. There is syntax that indicates that a regular expression is to be repeated some number of times and this is discussed in more detail in Section 5.3.1.3

Regular expressions are constructed analogously to arithmetic expressions by using various operators to combine smaller expressions. Concatenating regular expressions yields a regular expression that matches any string formed by concatenating strings that match the concatenated subexpressions. Of some specific interest is alternation using the | operator, quantifiers that determine how many times a construct may be applied (see below), and grouping of regular expressions using brackets (parentheses), (). For example, the regular expression (foo|bar) matches either the string foo or the string bar. The precedence order of the operations is that repetition is highest, then concatenation and then alternation. Enclosing specific subexpressions in parentheses overrides these precedence rules.

5.3.1.1 Character classes

A *character class* is a list of characters listed between square brackets, [and], and it matches any single character in that list. If a caret, ^, is the first character of the list, then the match is to any character **not** in the list. For example, [AGCT] matches any one of A, G, C or T, while [^123] matches any character that is not a 1, 2 or 3. A range of characters may be specified by giving the first and last characters, separated by a dash, such as [1-9], which represents all single digits between 1 and 9. Character ranges are interpreted in the collation order of the current locale. The following rules apply to metacharacters that are used in a character class: a literal] can be included by placing it first; a literal ^ can be included by placing it anywhere but first; a literal -, must be placed either first or last. Alternation does not work inside character classes because | has its literal meaning.

The period . matches any single character except a new line, and is sometimes referred to as the *wild card* character. Special shorthand notation for different sets of characters are often available; for example, \d represents any decimal digit, \s is shorthand for any space character, and their upper-case versions represent their negation. The symbol \w is a synonym for [[:alnum:]_], the alphanumeric characters plus the underscore, and \W is its negation.

Exercise 5.8
Write a function that takes a character vector as input and checks to see which elements have only nucleotide characters in them.

The set of POSIX character classes is given in Table 5.1. These POSIX character classes only have their special interpretation within a regular expression character class; for example, [[:alpha:]] is the same as [A-Za-z].

5.3.1.2 Anchors, lookaheads and backreferences

An *anchor* does not match any specific character, but rather matches a position within the text string, such as a word boundary, a place between characters, or the location where a regular expression matches. Anchors are zero-width matches. The symbols \< and \>, respectively, match the empty string at the beginning and end of a word. In the example below, we use gregexpr to show all the beginnings and endings of words. Notice that the length of the match is always zero.

[:alnum:]	alphanumeric characters: [:alpha:] and [:digit:].
[:alpha:]	alphabetic characters: [:lower:] and [:upper:].
[:blank:]	blank characters, space and tab.
[:cntrl:]	control characters. In ASCII, these characters have octal codes 000 through 037, and 177 (DEL).
[:digit:]	the digits: 0 1 2 3 4 5 6 7 8 9.
[:graph:]	graphical characters: [:alnum:] and [:punct:].
[:lower:]	lower-case letters in the current locale.
[:print:]	printable characters: [:alnum:], [:punct:] and space.
[:punct:]	punctuation characters: ^ ! " # $ % & ' ( ) * + , − . / : ; < = > ? @ [ ] \ _ {	} * and ~
[:space:]	space characters: tab, newline, vertical tab, form feed, carriage return, and space.
[:upper:]	upper-case letters in the current locale.
[:xdigit:]	hexadecimal digits: 0 1 2 3 4 5 6 7 8 9 A B C D E F a b c d e f.

Table 5.1: Predefined, POSIX, character classes.

```
> gregexpr("\\<", "my first anchor")

[[1]]
[1]  1  4 10
attr(,"match.length")
[1] 0 0 0

> gregexpr("\\>", "my first anchor")

[[1]]
[1]  3  9 16
attr(,"match.length")
[1] 0 0 0
```

The caret ^ and the dollar sign $ are meta-characters that, respectively, match at the beginning and end of a line. The symbol \b matches the empty string at the edge of a word (either the start or the end); and \B matches the empty string provided it is not at the edge of a word. In the code below, we show that \b is equivalent to both \> and \<, and that ^ and $ match only once per string.

```
> gregexpr("\\b", "once upon a time")
```

```
[[1]]
[1]  1  5  6 10 11 12 13 17
attr(,"match.length")
[1] 0 0 0 0 0 0 0 0

> gregexpr("\\>", "once upon a time")

[[1]]
[1]  5 10 12 17
attr(,"match.length")
[1] 0 0 0 0

> gregexpr("\\<", "once upon a time")

[[1]]
[1]  1  6 11 13
attr(,"match.length")
[1] 0 0 0 0

> gregexpr("^", "once upon a time")

[[1]]
[1] 1
attr(,"match.length")
[1] 0

> gregexpr("$", "once upon a time")

[[1]]
[1] 17
attr(,"match.length")
[1] 0
```

The notion of lookaheads and lookbehinds was introduced in Perl 5, and to use them, you will need to set `perl=TRUE` in the function calls. One of the problems that they can help to solve is the problem of finding one specified regular expression that is not followed by another specified regular expression. The syntax is (?...) for matching lookahead, (?!...), for negative lookahead, (?<=...) for lookbehind, and (?<!...) for negative lookbehind. Implementations differ but it is easier to deal with lookahead, so it tends to allow more general regular expressions than lookbehind.

If you consider the problem of finding some letter, say an r, that is not followed by an r, then the solution without lookaheads is a bit more convoluted than with them. The problem is detecting an r at the end of a string since in that case, the usual regular expression does not match; as we see in the code

below, the first two lines use the regular expression r[^r], but as we see it
fails in the second example, where r is the last character, and this is because
the regular expression [^r] has to match something. With a lookahead, we
find both matches.

```
> regexpr("r[^r]", "asffrb", perl = TRUE)

[1] 5
attr(,"match.length")
[1] 2

> regexpr("r[^r]", "asffr", perl = TRUE)

[1] -1
attr(,"match.length")
[1] -1

> regexpr("r(?!r)", "asffrb", perl = TRUE)

[1] 5
attr(,"match.length")
[1] 1

> regexpr("r(?!r)", "asffr", perl = TRUE)

[1] 5
attr(,"match.length")
[1] 1
```

There are, of course, other ways to solve this problem, usually involving
alternation; for example, r[^r]|r$ would also solve the problem as stated.

The backreference \N, where N is a single digit, matches the substring previously matched by the Nth parenthesized subexpression of the regular expression. So, for example, this regular expression would find pairs of upper-case
letters ([A-Z])\1. While this problem (finding pairs of letters) can be solved
in other ways, the use of backreferences makes it particularly simple.

```
> gregexpr("([A-Z])\\1", "ABBBZZ")

[[1]]
[1] 2 5
attr(,"match.length")
[1] 2 2
```

?	The preceding item is optional and will be matched at most once.
*	The preceding item will be matched zero or more times.
+	The preceding item will be matched one or more times.
{n}	The preceding item is matched exactly n times.
'{n,}'	The preceding item is matched n or more times.
'{n,m}'	The preceding item is matched at least n times, but not more than m times.

Table 5.2: Repetition operators for regular expressions.

5.3.1.3 Quantifiers

While matching specific patterns is often all that is needed, there are many cases where special handling of repeated instances of the regular expression is useful. For example, one can identify a white-space character using [:blank:], but sometimes you want to identify all contiguous white-space characters. The Classical problems involve finding either one or none of something, finding at least one of something and so on. To address these problems, there are several repetition quantifiers. The repetition quantifier comes after the regular expression. Some of the more common methods of specifying repetition are given in Table 5.2. Repetition is greedy, so the maximal possible number of repeats is used.

One minor difference between PCRE and extended regular expressions is that if a quantifier is followed by a ?, then in PCRE the matching is not greedy. The difference is demonstrated in the example below. In the first call to regexpr, the match is to five characters, while in the second it is only three characters long.

```
> regexpr("AB{2,4}?", "ABBBBB")

[1] 1
attr(,"match.length")
[1] 5

> regexpr("AB{2,4}?", "ABBBBB", perl = T)

[1] 1
attr(,"match.length")
[1] 3
```

5.3.2   Matching

For each regular expression, each input sequence is traversed from left to right; when a match is found, the regular expression engine returns. In R, there are functions that will find all matches, and either report them or perform substitution on them. However, the convention that has been adopted in R (borrowed from Perl) is that no overlapping matches are detected. This is somewhat problematic for some biological problems, such as finding transcription factor binding sites, as these often overlap.

Different implementations may return different matches to the same regular expression. Typical differences arise due to whether a longer sequence has preference over a shorter one, as is shown in the example below. In the first call using extended regular expressions, the match is to `foobar`, while in the second call, using PCRE, the match is to the left-most query string.

```
> regexpr("foo|foobar", "myfoobar")

[1] 3
attr(,"match.length")
[1] 6

> regexpr("foo|foobar", "myfoobar", perl = TRUE)

[1] 3
attr(,"match.length")
[1] 3
```

Another problem that arises with application to biological sequence matching is shown by the convention used by `gregexpr`, which finds all non-overlapping matches to the input regular expression. In the example below, `gregexpr` reports only one match, since the second match begins at position 7, which is inside the first match. On the other hand, `gregexpr2` from the **Biostrings** package reports both. The current version of `gregexpr2` only supports exact matching and does not support any form of regular expression matching.

```
> testS = "ACTACCACTACCACT"
> gregexpr("ACTACCACT", testS)

[[1]]
[1] 1
attr(,"match.length")
[1] 9

> gregexpr2("ACTACCACT", testS)
```

```
[[1]]
[1] 1 7
```

5.3.3   Using regular expressions

We now consider a few examples; the first is adapted from Stubblebine (2007). They suggest using ^\d\d\/\d\d/\d\d\d\d$ to match the MM/DD/YYYY format. But that is not quite specific enough, since months can only range from 1 to 12 and days from 1 to 31. In the first example, the date is fine, but in the second, both the month and the day are not valid, and it might be nice to check that too.

```
> regexpr("\\d\\d\\/\\d\\d\\/\\d\\d\\d\\d",
          "today is 12/01/1977", perl = TRUE)

[1] 10
attr(,"match.length")
[1] 10

> regexpr("\\d\\d\\/\\d\\d\\/\\d\\d\\d\\d",
          "today is 21/41/1977", perl = TRUE)

[1] 10
attr(,"match.length")
[1] 10
```

Exercise 5.9
Create a valid regular expression that checks to make sure that both the month and day specifications are correct.

Our next example is a small function that strips leading or trailing white space from its input value.

```
> strwhite = function(x, lead = TRUE, trail = TRUE) {
      if (lead)
          x = sub("^[[:blank:]]*", "", x, perl = TRUE)
      if (trail)
          sub("[[:blank:]]*$", "", x, perl = TRUE)
      else x
  }
```

Exercise 5.10
*What is the purpose of the * in the regular expressions? Can you extend this
to deal with white space as defined by* [:space:]*? Write a function similar to*
strwhite *that replaces two or more leading blanks with a single space. Modify*
strwhite *to also strip* \n *from the end of a line.*

Regular expressions, although using a different syntax, have been used by
the Prosite database to describe protein motifs. An example of a Prosite motif is given below. The *pattern* for this motif is given on the line that begins
with PA, and is [RK]-x(2,3)-[DE]-x(2,3)-Y. The syntax for a Prosite regular expression is that either x or X matches any amino acid, the dash - is a
separator and has no meaning, square brackets contain lists of characters to
match and curly braces contain lists of characters that cannot match. Repetition is specified using brackets (,). The one-argument version specifies the
number of repetitions; with the two-argument version, the minimum number
is specified first, the maximum number second. The period at the end of the
Prosite regular expression is ignored.

```
ID    TYR_PHOSPHO_SITE; PATTERN.
AC    PS00007;
DT    APR-1990 (CREATED); APR-1990 (DATA UPDATE); APR-1990
      (INFO UPDATE).
DE    Tyrosine kinase phosphorylation site.
PA    [RK]-x(2,3)-[DE]-x(2,3)-Y.
CC    /TAXO-RANGE=??E?V;
CC    /SITE=5,phosphorylation;
CC    /SKIP-FLAG=TRUE;
CC    /VERSION=1;
DO    PDOC00007;
//
```

A Prosite pattern can be turned into a regular expression quite simply using
gsub, as we show in the code below. We first strip out the dashes and the
period that indicates the end of the pattern. Next, we translate the Prosite
wild card character to the regular expression wild card character and change
the brackets that surround the repetition indicator.

```
> prositeM = "[RK]-x(2,3)-[DE]-x(2,3)-Y."
> regexM = gsub("-|\\.", "", prositeM)
> regexM = chartr("xX()", "..{}", regexM)
```

And in the code example below, we test whether our translation worked.

```
> testP = "ACRDRACDTUYACRD"
> testN = "ACRDRAXXCDTUYACRD"
> regexpr(regexM, testP)

[1] 5
attr(,"match.length")
[1] 7

> regexpr(regexM, testN)

[1] -1
attr(,"match.length")
[1] -1
```

5.3.4   Globbing and regular expressions

On Unix-like systems, *globbing* expands file names using a pattern-matching notation similar to that of regular expressions. However, the capabilities are much more limited, and the uses are typcially for finding files with particular endings, e.g., `ls *.pdf` to list all files that end with `.pdf`. The `glob2rx` function translates globbing patterns into corresponding regular expressions.

5.4   Prefixes, suffixes and substrings

There are a number of classical string finding problems that have broad application. They are the problems of finding the longest common prefix, suffix or substring across a set of strings. A related problem is that of finding the longest repeated string in a series of strings, where here that repeated string could be entirely within one of the set of strings, or it could be in two or more.

The first two problems are quite easy to solve, as one simply starts at one end of the strings, or the other, and compares characters until they do not match. There are functions in the **Biobase** package, `lcPrefix` and `lcSuffix`, that address the suffix and prefix problems. But the longest common substring problem, and the longest repeated substring problems, are much harder, and one elegant solution makes use of a data structure known as a suffix tree. Suffix trees are discussed in many places, such as Cormen et al. (1990) and Gusfield (1997); among the more interesting approaches is that in Chapter 15 of Bentley (1999). Suffix trees are widely used in Bioinformatics and underly the MUMmer technology (Kurtz et al., 2004).

In the code below, we demonstrate the use of the suffix and prefix functions. Notice that white space is considered to be part of the prefix or suffix.

```
> library("Biobase")
> str1 = c("not now", "not as hard as wow", "not something new")
> lcPrefix(str1)

[1] "not "

> lcSuffix(str1)

[1] "w"
```

For any string, then, the set of suffixes is of the same length as the string; the first suffix is the whole string, the second suffix is the string starting at the second letter, and so on. Consider the work *biology*, which has seven characters, and hence seven suffixes:

```
1] biology
2] iology
3] ology
4] logy
5] ogy
6] gy
7] y
```

And these can then be sorted into lexicographic order to yield:

```
1] biology
2] iology
3] gy
4] logy
5] ology
6] ogy
7] y
```

And now the longest repeated subsequence, or substring, can be found by comparing adjacent pairs of suffixes. In this case it is very short; the letter o appears twice. The package **Rlibstree**, available from the Omegahat Project, provides some tools for computing with suffix trees.

```
> library("Rlibstree")
> s1 = "biology"
> getLongestSubstring(s1)
[1] "o"
```

5.5 Biological sequences

The genome of every organism is encoded in chromosomes that consist of either DNA or RNA. High throughput sequencing technology has made it possible to determine the sequence of the genome for virtually any organism, and there are many that are currently available. Chromosomes for many organisms can be thought of as very long strings from a relatively small alphabet. DNA is a double-stranded molecule, where bases on opposite strands are complementary, in that A is complementary to T, and C is complementary to G. The RNA alphabet is very similar, with U representing uracil, which is found in RNA but not DNA. However, in many cases, either the exact nucleotide at any location is unknown, or is variable, and the International Union of Pure and Applied Chemistry (IUPAC) has provided a standard nomenclature suitable for representing such sequences. The alphabet for dealing with protein sequences is based on the 20 amino acids.

The discussion here is based on code provided in the **Biostrings** package. The basic class used to hold strings is the *BString* class, which has been designed to be efficient in its handling of large character strings. Subclasses include *DNAString*, *RNAString* and *AAString* (for holding amino acid sequences). The *BStringViews* class holds a set of *views* on a single *BString* instance; each view is essentially a substring of the underlying *BString* instance. Alignments are stored using the *BStringAlign* class.

Fundamental operations, and the corresponding **Biostrings** functions, on DNA and RNA sequences are listed next.

complement replace each base in the input string with its complementary base.

reverse return a string with the bases in the reverse order.

reverseComplement both reverse and complement the input string.

transcribe given an input DNA sequence, return the value of the RNA sequence that would result from transcribing the input.

cDNA given an input RNA sequence, return the complementary DNA (cDNA) sequence that gave rise to it.

In the example below, we begin with the RNA sequence for a human microRNA and determine the DNA sequence that gave rise to it.

```
> st1 = RNAString("UCUCCCAACCCUUGUACCAGUG")
> cD = cDNA(st1)
```

The **matchprobes** package was written primarily to deal with Affymetrix GeneChips and hence focuses on them. The functions provided include `basecontent`, `complementSeq` and `reverseSeq`. In the code below, we demonstrate one method for obtaining the mismatch probe.

```
>   library("matchprobes")
>   seq <- c("CGACTGAGACCAAGACCTACAACAG",
            "CCCGCATCATCTTTCCTGTGCTCTT")
>  complementSeq(seq, start=13, stop=13)

[1] "CGACTGAGACCATGACCTACAACAG" "CCCGCATCATCTATCCTGTGCTCTT"
```

Exercise 5.11
Write a version of `complementSeq` *that works for either DNA or RNA using* `chartr`. *How does the speed compare with that of the version in the* **matchprobes** *package? Write a version of* `reverseSeq` *using* `strsplit`, `rev` *and* `paste`. *How does the speed of that function compare with the one in the* **matchprobes** *package?*

5.5.1   Encoding genomes

A number of complete genomes, represented as *DNAString* objects, are provided through the Bioconductor Project. They rely on infrastructure in the **BSgenome** package, and all such packages have names that begin with BSgenome. You can find the list of available genomes using the `available.genomes` function. In the code below, we load build 18 of the human genome, and show what data are contained in the package.

```
> library("BSgenome.Hsapiens.UCSC.hg18")
> Hsapiens


Human genome
|
| organism: Homo sapiens
| provider: UCSC
| provider version: hg18
| release date: Mar. 2006
| release name: NCBI Build 36.1
|
| single sequences (see '?seqnames'):
|    chr1            chr2            chr3            chr4
|    chr5            chr6            chr7            chr8
```

```
|   chr9            chr10           chr11            chr12
|   chr13           chr14           chr15            chr16
|   chr17           chr18           chr19            chr20
|   chr21           chr22           chrX             chrY
|   chrM            chr5_h2_hap1    chr6_cox_hap1    chr6_qbl_hap2
|   chr1_random     chr2_random     chr3_random      chr4_random
|   chr5_random     chr6_random     chr7_random      chr8_random
|   chr9_random     chr10_random    chr11_random     chr13_random
|   chr15_random    chr16_random    chr17_random     chr18_random
|   chr19_random    chr21_random    chr22_random     chrX_random
|
| multiple sequences (see '?mseqnames'):
|   upstream1000  upstream2000  upstream5000
|
| (use the '$' or '[[' operator to access a given sequence)
```

As you see in the output, all chromosomes are present, as are other pieces. In many genomic sequences, there are regions that are known not to be be of interest for a specific task. These include regions where the sequence is unknown (coded as an N in genomic sequences), or regions with short repeats. These can be *masked* using the mask function, which will mask features based either on their position or on their content. The basic idea used is to create a view on the original string that only contains the regions that are not masked. In the example below, we mask all Ns on human Chromosome 22; we also report the proportion of Ns.

```
> chr22NoN <- mask(Hsapiens$chr22, "N")
> alphabetFrequency(Hsapiens$chr22, freq = TRUE)["N"]

    N
0.299
```

5.6 Matching patterns

The **Biostrings** package provides three basic matching methods. One method does exact matching of a single query sequence against a single reference sequence (matchPattern); a second matches patterns that are of the form left-gap-right (matchLRPattern), allowing for different numbers of mismatches in the left and right patterns, and for specifying the maximum number of characters in the gap. The third method (`matchPDict`) compares a large library of query sequences to a single reference sequence. It supports different length query sequences and both exact and inexact matching. Extensions and improvements are planned, so it is important to read the documentation for the **Biostrings** package to determine what the current capabilities are.

5.6.1 Matching single query sequences

A motif is a short sequence pattern that occurs repeatedly in a group of related DNA or RNA sequences or that occurs in protein or peptide sequences. The existence of a motif is suggestive of a conserved function. For DNA and RNA, motifs are often indicators of promotor binding sites (the famous TATA box) or of transcription factor binding sites, or of splicing signals. In protein sequences, motifs usually reflect structural or functional conservation.

Among the more famous motifs in DNA sequences is the so-called TATA box consensus sequence (TATAAAA), which is involved in guiding RNA polymerase II to the initiation site for transcription. One can either search for exact matches, as in the code below, or for some number of mismatches. If you only want to know how many matches there are, not where they are, then use the `countPattern` function.

```
> TATA = "TATAAAA"
> mT = matchPattern(TATA, chr22NoN)
> countPattern(TATA, chr22NoN)

[1] 5276
```

While one might expect that matching to the masked version of Chromosome 22 would be faster, this need not be the case. The issue is primarily due to the current implementation where a masked sequence is just a set of views on the original sequences and `matchPattern` is called on each view in an R `for` loop. The cost of this surpasses the benefit that you get from reducing the length of the target sequence. Using `chr22NoN` is reasonable because the number of views is small, but with a very fragmented masked sequence (thousands of views), things would be much worse.

Typically one might be willing to live with some number of mismatches, and that too can be accommodated using `matchPattern` (although the function becomes appreciably slower as more mismatches are allowed). In the code below, we also demonstrate the use of the `mismatch` function that shows the location of the mismatch(es) in each of the patterns. The value is a zero length integer vector if there are no mismatches.

```
> mmT = matchPattern(TATA, chr22NoN, max.mismatch = 1)
> length(mmT)

[1] 102104

> mismatch(TATA, mmT[1:3])

[[1]]
[1] 2

[[2]]
[1] 5

[[3]]
[1] 7
```

5.6.2 Matching many query sequences

Matching a huge number of query sequences to a single target sequence is a problem that is now relevant due to high throughput sequencing technologies. These technologies typically yield a large number, sometimes in the tens of millions, of short *reads*. One of the bioinformatic tasks is to match these to a known genome. And the function `matchPDict` can be used for this. It is based on the Aho-Corasick algorithm.

The following example is taken from the `matchPDict` manual page. Except here, we match probes from the Affymetrix HG-U95Av2 GeneChip to Chromosome 22. First the library containing the probe information is loaded, then we create a dictionary (preprocess the approximately 200,000 25-mers), and finally match that to Chromosome 22. One might want to also search for the mismatch probes, which are not stored in the probe packages since they are easily obtained by taking each probe and replacing its 13th nucleotide with its complement; this can easily be achieved with the `complementSeq` function from the **matchprobes** package.

```
> library(hgu95av2probe)
> dict <- hgu95av2probe$sequence
> length(dict)

[1] 201800

> unique(nchar(dict))

[1] 25
```

```
> dict[1:5]

[1] "TGGCTCCTGCTGAGGTCCCCTTTCC" "GGCTGTGAATTCCTGTACATATTTC"
[3] "GCTTCAATTCCATTATGTTTTAATG" "GCCGTTTGACAGAGCATGCTCTGCG"
[5] "TGACAGAGCATGCTCTGCGTTGTTG"

> pdict <- PDict(dict)
> vindex <- matchPDict(pdict, Hsapiens$chr22)
> length(vindex)

[1] 201800

> count_index <- countIndex(vindex)
> sum(count_index)

[1] 53280

> table(count_index)

count_index
     0      1      2      3      4      5      6      7
198516   2855    185     84     56     12      6      6
     8     10     11     13     15     35     52     68
     3      1      2      1      1      1      1      1
    90    147    152    179    186    190    194    196
     1      1      1      1      1      1      1      1
   197    205    214    249    264    274    283    289
     1      1      1      1      1      1      1      1
   297    309    310    324    330    333    335    338
     1      1      1      1      1      1      1      1
   365    384    413    417    421    444    453    460
     2      1      1      1      1      1      1      1
   461    467    479    486    492    502    514    517
     1      1      2      1      1      1      1      1
   788    823    857    886    904    921    932    953
     1      1      1      1      1      1      1      1
   973   1146   1147   1173   1206   1227   1269   1270
     1      1      1      1      1      1      1      1
  1283   1297   1299   1303   1305   1309   1315   1957
     1      1      1      1      1      1      1      1
  2127   2757   2771
     1      1      1
```

Most of the 25-mers do not match at all, but some match a very large number of times, suggesting that they are not that specific. Note that we have only matched to one strand of Chromosome 22, and expect most 25-mers to match at some other location in the genome. We can confirm our results using `countPattern`, as is shown in the example below.

```
> dict[count_index == max(count_index)]

[1] "CTGTAATCCCAGCACTTTGGGAGGC"

> countPattern("CTGTAATCCCAGCACTTTGGGAGGC", Hsapiens$chr22)

[1] 2771
```

The functions `startIndex` and `endIndex` get the starting and ending indices, respectively.

5.6.3 Palindromes and paired matches

Palindromes are words or sequences of characters that read the same forward as they do backward, such as the word *madam*. While there are natural language applications, finding palindromes, or sequences like palindromes, have important biological applications. We extend the definition of palindrome a little so that it is more relevant. The variants we are interested in are composed of a left arm of the palindrome, a *loop* of some number of characters, followed by the right arm of the palindrome. In addition there will be cases where we would like to find left and right arms that are reverse complements of each other (and hence may hybridize). If the loop is zero, then these are the *complemented palindromes* defined in Gusfield (1997), but the loop plays an important role in some applications. The **Biostrings** package has two functions for finding palindromes: `findPalindromes` and `findComplementedPalindromes`. The latter can only be used on sequences such as DNA or RNA where the notion of complement is sensible.

The example below is based on the manual page for `findPalindromes`. However, we use human chromosome 22.

```
> chr22_pals = findPalindromes(chr22NoN, min.armlength = 40,
                        max.looplength = 20)
> nchar(chr22_pals)

 [1]  83  96 107  94  81  90  88  91  88 136  91  88 106 100
[15] 100  88  97  81  82  81  85  89  93  97 101 105 109 111
[29] 107 103  99  95  91  87  83  81  86  83  85  91  83  89
[43] 113  83  96  98  97 127  95  80  85  88  83 100  97  94
[57]  87  83 105 104  81  83  93
```

178   R Programming for Bioinformatics

```
> palindromeArmLength(chr22_pals)

 [1]  83  96 107  94  81  43  88  40  40  64  40  40 106  43
[15]  43  40  40  81  82  81  85  89  93  97 101 105 109 111
[29] 107 103  99  95  91  87  83  81  86  83  85  91  83  89
[43]  47  83  96  98  97 127  95  80  85  88  83  40  40  41
[57]  41  83 105 104  81  83  93

> palindromeLeftArm(chr22_pals)

  Views on a 49691432-letter DNAString subject
subject: NNNNNNNNNNNNNNNNNNNNNNNNNNNN...NNNNNNNNNNNNNNNNNNNNNNNNNNNN
views:
         start       end width
 [1] 14668595 14668677    83 [GCGCGTGCGGCGTG...GTGCGGCGTGCGCG]
 [2] 14668595 14668690    96 [GCGCGTGCGGCGTG...GTGCGGCGTGCGCG]
 [3] 14668596 14668702   107 [CGCGTGCGGCGTGC...CGTGCGGCGTGCGC]
 [4] 14668609 14668702    94 [CGCGTGCGGCGTGC...CGTGCGGCGTGCGC]
 [5] 14668622 14668702    81 [CGCGTGCGGCGTGC...CGTGCGGCGTGCGC]
 [6] 17158213 17158255    43 [ATAGATAGATAGAT...TAGATAGATAGATA]
 [7] 17158216 17158303    88 [GATAGATAGATAGA...AGATAGATAGATAG]
 [8] 18530077 18530116    40 [ACCACCACCACCAC...CACCACCACCACCA]
 [9] 18530368 18530407    40 [ACCACCACCACCAC...CACCACCACCACCA]
  ...      ...       ...   ... ...
[55] 35557739 35557778    40 [TTCTTCTTCTTCTT...CTTCTTCTTCTTCT]
[56] 35557742 35557782    41 [TTCTTCTTCTTCTT...TTCTTCTTCTTCTC]
[57] 36054993 36055033    41 [CTCTCTCTCTCTCC...TCTCTCTCTCTCTC]
[58] 40665769 40665851    83 [AAAGAAAGAAAGAA...AAGAAAGAAAGAAA]
[59] 41828847 41828951   105 [ATATACATATATAC...CATATATACATATA]
[60] 42640359 42640462   104 [CTTCTTCTTCCTTC...CTTCCTTCTTCTTC]
[61] 47175937 47176017    81 [AAGAAAGAAAGAAA...AAAGAAAGAAAGAA]
[62] 47175938 47176020    83 [AGAAAGAAAGAAAG...GAAAGAAAGAAAGA]
[63] 47948015 47948107    93 [TATATATATATATA...ATATATATATATAT]

> ans = alphabetFrequency(chr22_pals,
        base = TRUE)
> head(ans, n = 15)

      A  C  G  T other
[1,]  0 26 45 12     0
[2,]  0 30 52 14     0
[3,]  0 34 57 16     0
[4,]  0 30 50 14     0
[5,]  0 26 43 12     0
[6,] 46  0 21 23     0
```

```
 [7,] 44  0 22 22      0
 [8,] 31 54  0  6      0
 [9,] 30 52  0  6      0
[10,] 46 80  0 10      0
[11,] 31 54  0  6      0
[12,] 30 52  0  6      0
[13,] 36 62  0  8      0
[14,] 33 59  0  8      0
[15,] 34 57  0  9      0
```

We see from the frequency counts that the palindromic regions have some-what unusual frequencies. Typically, one base is not present at all, and the other bases form some sort of repeated sequences.

Exercise 5.12
Find all of the palindromes that have all four bases present. Are their sequences also highly repetitive?

Exercise 5.13
Find all the complemented palindromes on Chromosome 22.

Palindromes are a special case of paired matches. For a paired match, one specifies a left pattern, a right pattern and a maximum distance between the left pattern and the right pattern. This type of matching is handled by the `matchLRPatterns` function.

```
> Lpattern <- "CTCCGAG"
> Rpattern <- "GTTCACA"
> LRans = matchLRPatterns(Lpattern, Rpattern, 500,
      Hsapiens$chr22)
> length(LRans)

[1] 21
```

And we see that there are 21 places on Chromosome 22 where the left pattern occurs within 500 bases of the right pattern.

5.6.4 Alignments

One of the major tasks in Bioinformatics is sequence alignment. Both Gus-field (1997) and Haubold and Wiehe (2006) give reasonable coverage of the problems, and the methods used to solve them. Our covarage is quite brief. There are many algorithms that can be used to perform alignments, and many different tasks, such as motif finding, aligning multiple sequences, aligning genes, aligning genomes, local versus global alignments, and all often require slightly different tools. The tools currently available in Bioconductor are limited and currently only support pairwise global alignment. There are two basic types of algorithms that are widely used: *optimal alignment* that typically relies on dynamic programming and *heuristic alignments* that do not necessarily provide an optimal solution but are fast and allow users to work on large problems.

Two, possibly related, biological sequences can differ in a number of ways. One typically thinks of the relatedness as coming from evolutionary time, where the sequences shared some common ancestor. The simple sorts of changes that can occur are point mutations, insertions and deletions (the latter two are referred to as *indels*). Choosing the optimal alignment involves some form of scoring; and for amino acid alignments, substitution matrices are used. For aligning DNA and RNA, one typically uses some score for mutations and another score for insertions or deletions, although modifications are somewhat straightforward. Most researchers believe that some form of affine penalty for indels is better than a simple penalty per nucleotide.

The alignment of protein sequences relies on a substitution matrix, which provides penalties for the different substitutions. A good discussion of the current methodology used to create these matrices is given in Eddy (2004). The example below shows how to align two amino acid sequences using the needwunsQS function, and one of the available substitution matrices (look for help on substitution.matrices to find all the predefined ones). It is important to emphasize that this implementation of the Needleman-Wunsch algorithm does not support the affine model for gap penalties; all gaps are treated the same.

```
> aa1 <- AAString("HXBLVYMGCHFDCXVBEHIKQZ")
> aa2 <- AAString("QRNYMYCFQCISGNEYKQN")
> needwunsQS(aa1, aa2, "BLOSUM62", gappen = 3)

Global Pairwise Alignment
1:   HXBLVYMGCHFDCXV-BEHIKQZ
2:   QRN--YMYC-FQCISGNEY-KQN
Score:   39

> needwunsQS(aa1, aa2, "BLOSUM62", gappen = 8)

Global Pairwise Alignment
1:   HXBLVYMGCHFDCXVBEHIKQZ
2:   QRN--YMYC-FQCISGNEYKQN
Score:   17
```

Aligning DNA is a slightly different problem; here the scoring is usually handled more simply with a negative score given for mismatches, and a gap penalty. In the example below, we load the sequence for two genes, of known homology in two different yeast strains, *S. cerevisae* and *S. paradoxus*; the standard name for the gene in *S. cerevisae* is YDL143W. In the second call to needwunsQS, we made the gap penalty larger than the mismatch penalty, with the consequence that the aligned sequence is now shorter, since it is more expensive to add a gap than it is to have a mismatch. The data are in the extdata directory of the **Biostrings** package, so we first read them in and then align them, using different gap penalites.

```
> oldD = setwd(system.file("extdata", package = "Biostrings"))
> Sc = readFASTA("Sc.fa", )[[1]]$seq
> Sp = readFASTA("Sp.fa")[[1]]$seq
> setwd(oldD)
> mat <- matrix(-5L, nrow = 4, ncol = 4)
> for (i in seq_len(4)) mat[i, i] <- 0L
> rownames(mat) <- colnames(mat) <- DNA_ALPHABET[1:4]
> dnaAlign1 = needwunsQS(Sc, Sp, mat, gappen = 1)
> nchar(dnaAlign1)

[1] 1704
```

In the example above, the gap penalty was much smaller than the mutation penalty, so gaps will be preferred over mutations. If we increase the gap penalty, so that gaps become more expensive, then mutations will be favored.

```
> dnaAlign2 = needwunsQS(Sc, Sp, mat, gappen = 6)
> nchar(dnaAlign2)

[1] 1587
```

Exercise 5.14
Over evolutionary time methylated cytosines (C) are converted to thymines (T) due to spontaneous deamination. Modify the penalty matrix mat *above to penalize less for this conversion than for the others. How does that change the two alignments?*

Some investigators want to find the longest subsequence that is common to both strings (this is sometimes referred to as the maximum unique match). This is easily handled using the **Rlibstree** package.

```
> library("Rlibstree")
> tree = SuffixTree(c(Sc, Sp))
> MUM = getLongestCommonSubstring(tree)
> nchar(MUM)

[1] 89
```

The consensus matrix for any alignment can be obtained using the `consmat` function. The consensus matrix is simply the matrix where rows correspond to characters in the alphabet, columns correspond to positions in the sequence, and for each column the proportion of each letter found in that position is reported. It is of somewhat limited value for pair-wise alignments, but we provide a brief example below.

```
> consmat(dnaAlign1)[, 1:20]

       pos
letter 1 2 3 4 5 6 7 8 9 10 11  12  13 14 15  16  17 18 19 20
     - 0 0 0 0 0 0 0 0 0  0  0 0.5 0.5  0  0 0.5 0.5  0  0  0
     A 1 0 0 0 0 0 0 0 0  1  1 0.0 0.5  0  0 0.0 0.0  0  0  1
     C 0 0 0 0 1 0 0 1 0  0  0 0.0 0.0  0  0 0.5 0.0  1  1  0
     G 0 0 1 0 0 0 1 0 0  0  0 0.5 0.0  1  0 0.0 0.0  0  0  0
     T 0 1 0 1 0 1 0 0 1  0  0 0.0 0.0  0  1 0.0 0.5  0  0  0
```

Chapter 6

Foreign Language Interfaces

6.1 Introduction

In this chapter we discuss some of the many different interfaces to functions and libraries written in other languages. There are several reasons for wanting to interact with software written in other languages. The two main reasons are efficiency and access to existing code bases. Since R is not compiled, in some situations its performance can be substantially improved by writing code in a compiled language. There are also reasons not to write code in other languages, and in particular we caution against premature optimization, prototyping in R is often cost effective. And in our experience very few routines need to be implemented in other languages for efficiency reasons. Another substantial reason not to use an implementation in some other language is increased complexity. The use of another language almost always results in higher maintenance costs and less stability. In addition, any extensions or enhancements of the code will require someone that is proficient in both R and the other language.

We focus most of our attention on writing and linking to C code since it is the most widely used interface and because it is the mechanism used to interface with other languages as well. In large part, the popularity of the C interface is due to the fact that R itself is largely written in C and it is easy to make use of the internal data structures, macros and code from routines written in C. We will briefly discuss FORTRAN, Perl, and Python, but their treatment is not in-depth and readers are referred to other sources, such as the R Extensions manual (R Development Core Team, 2007c) and the documentation supplied with a package that they want to use.

The organization of this chapter, after a brief overview, follows the basic tasks that a programmer will need to carry out to successfully use software written in a foreign language. They will need to write the R code, write the C code, correctly compile and link the C code, ensure that a library is placed in an appropriate location and that the correct entry point is found when the code in the R function is evaluated in R. These tasks can be quite complex and we recommend that developers who access routines written in foreign languages place that code in a package; see Chapter 7 for details on how to write R packages. There is substantial support in the package building and loading mechanism for using calls to C and other languages. It is possible to call directly to appropriately built libraries but this approach tends to be quite cumbersome, and we do not discuss it.

6.1.1   Overview

We begin with a brief overview of the basic tasks that must be carried out in order to use software written in another language from within R. The basic set of operations that must be carried out are reasonably well known. There must be some mechanism for finding the the appropriate software entry point, there must be some mechanism to translate R data structures into data structures used by the language being invoked, there must be some mechanism to initiate the call, and finally there must be some mechanism for getting return values and passing them back to R. It is helpful if there is also a mechanism for detecting and handling exceptions such as errors.

Perhaps the most important thing to realize is that when interacting with code written in another language, there must be some translation of data structures from one language to the other. Any data structure that can be translated can be passed from one language to the other. When using the .C interface, for example, there is a one-to-one correspondence between data representations in R and C; see Table 6.1 for more details. And in this case the developer has the option of either copying the data or not.

Some attention should also be paid to issues of memory management. R has a sophisticated internal garbage collection system, which can be used for many programming problems. But working with the memory manager requires that the program follow the appropriate paradigms. On the other hand, using language-specific calls to allocate memory (e.g., `malloc` in C) can impose a substantial burden on the developer. Forgetting to free memory can lead to memory leaks, and such purpose-allocated memory cannot easily be used for variables that are passed back to R.

In general, one does not call the interface functions such as .C or .Call directly, but rather embeds these calls inside larger functions that set up the appropriate calling sequence, invoke the foreign function and then process the return values.

The foreign function itself is contained in some form of dynamic library that will be loaded into R, either directly by a call to `dyn.load` or when the package containing the C code is loaded. When one of these functions is called, R looks through the set of dynamically loaded libraries for one with the appropriate symbol and invokes it. Since it is not uncommon for two authors to use similar names for functions, there are real chances for inadvertently invoking the wrong C routine. Using R packages for any external libraries and the use of the registration mechanism ensure that the correct code is called.

It is important when writing functions that you take advantage of as much of the internal code in R as possible. The R API is discussed in Section 6.4. There are many functions for creating R objects and for duplicating R objects.

Other functions, such as the internal versions of subscripting, sorting and, most importantly, random number generation, can be called directly in your code. We strongly recommend that you make use of as much of the internal code as possible.

6.1.2   The C programming language

We presume that the reader is familiar with programming in C. If not, there are many good books that describe C; perhaps the best and most widely used is Kernighan and Ritchie (1988). There are also a number of interesting books with algorithms and code examples written in C. Among these are Sedgewick (2001), which is a very good reference.

In C, definitions and declarations that need to be shared between files are declared in header files. Header files are then included in each file that requires access to the definitions. Header files have a `.h` suffix. When interfacing with R, you will generally need to include some of the R header files so that your code can make use of internal R data structures and functions. These are described in Section 6.4.1.

The C language supports pointers, and these are used extensively in the internals of R, as are macros and structures. These various topics are well described in Kernighan and Ritchie (1988), and readers unfamiliar with them are advised to spend some time studying them.

6.2   Calling C and FORTRAN from R

There are a number of different interfaces that can be used to call C from R. They include `.Call`, `.External`, and `.C`. The first pass down pointers to internal R objects, which can then be manipulated in C. But that requires that the C code be written using R internal data structures, and familiarity with the R internals will be essential. The other interface, `.C`, passes down pointers to C data structures and is suitable for calling C code that is not aware of R's internal data structures. Details on the specific conversions are listed in Table 6.1. These interfaces also differ in their return values. `.C` and `.Fortran` do not return any value, but rather rely on altering the values that were passed out and having the calling function process them on return. For `.Call` and `.External`, the return value is an R object (the C functions must return a `SEXP`), and for these functions the values that were passed are typically not modified. If they must be modified, then making a copy in R, prior to invoking the C code, is necessary.

The first argument to all interface functions is `name`, which is the name of the foreign function to be invoked. This is the name of the C routine as it appears in your C code (or the name of the FORTRAN subroutine). Some compilers prepend underscores or perform other operations on the names of functions (*name-mangling*). There is no need for you to perform any name-mangling. R detects when and how the names will be mangled, for each platform, and performs the name-mangling automatically.

The second argument to all interface functions is the ... argument and the values supplied here are passed down to the function being called. The order in this list is the order in which they will be passed to the external function. Any names in the list are ignored during the call to .C or .Fortran, but can be used to extract components of the returned value.

For functions within a package with a name space, the PACKAGE argument to all foreign interface functions should be omitted. This will ensure that the dynamic-link library (DLL) for the correct version of the package is found and used.

So, a first design decision is whether to pass down basic C data structures, or whether more complex objects, such as lists, or instances of classes, etc., will be needed at the C level. If the former, then using .C will be the easiest approach. We note that this is also a reasonable approach when developing an interface to an existing set of C functions. Alternatively, one can write a small interface routine and use .Call.

There are a number of advantages to using the .Call interface that by far outweigh the slightly increased complexity of use. Among these is the fact that it is much easier to check and ensure that the types, lengths (or sizes), etc., of the arguments are correct, thereby providing more security against errors at the C level. Since R data objects are largely self-describing, they can be queried in the C code so an error can be thrown if a problem is encountered. By contrast, the .C interface essentially passes pointers to C data structures, which are not self-describing. Thus the external C code must presume that the input values are of the correct size and type, since no direct validation is possible.

There are some examples in the package **RBioinf** that accompanies this monograph. In particular, simpleRand and simpleSort demonstrate some uses of the .Call interface, and the implementation is discussed in some detail in Section 6.4.

6.2.1   .C and .Fortran

The first argument to these functions is the name of the routine to be called, as it appears in the source file. The next formal argument is the ... argument, and values passed via the ... arguments are supplied to the native routine, in the order given. Values can be named anything other than the four names discussed in the next paragraph, but the names are ignored in the call to the foreign function. There is a limit of 65 arguments that can be passed to a native routine. If a value is named, then the names can be used to extract the relevant components from the return value.

R Storage Mode	C Type	FORTRAN Type	SEXPTYPE
logical	int *	INTEGER	LGLSXP
integer	int *	INTEGER	INTSXP
double	double *	DOUBLE PRECISION	REALSXP
single	single *	SINGLE PRECISION	SINGLESXP
complex	Rcomplex *	DOUBLE COMPLEX	CPLXSXP
character	char **	CHARACTER*255	STRSXP
raw	char *	none	RAWSXP
list	SEXP	not supported	VECSXP
other	SEXP	not supported	none

Table 6.1: Type conversions between internal R types and C and FORTRAN types. Note that the single type only exists in R as a means to pass values out to C or FORTRAN.

The calls to both .C and .Fortran include four arguments: DUP, NAOK, PACKAGE and ENCODING. These must be named arguments in the call, and no partial matching is used. DUP controls whether or not the arguments are copied before being passed to the external function. Not copying is generally done for efficiency reasons but can be dangerous. Since nothing is returned from these functions, they must modify at least one of the supplied arguments. It is somewhat safer to modify copies of objects, as that ensures that internal R data objects are not altered. If NAOK is set to TRUE, then missing values, NAs, and other non-finite values, such as Inf, are passed out to the external function; otherwise an error is signaled. NAOK should be set to FALSE when calling routines that were not specifically written to interface with R, since it is unlikely that they will appropriately accommodate these special values. The PACKAGE argument specifies the name of the package and limits the search for the function to the DLL that is associated with the named package. The ENCODING argument can be used to provide encoding information for character data that is being passed to the external code.

6.2.2 Using .Call and .External

These functions provide access to C code that is R-aware, and hence their calling sequence is much simpler. The first argument is the name of the C function that will be invoked, followed by the ... argument and then the PACKAGE argument. Both the ... argument and the PACKAGE argument behave as described above. Neither .Call nor .External copy their arguments, so values passed down through these interfaces should be treated as read-only; altering them will be reflected in the corresponding values at the R level and hence violate the pass-by-value semantics of the language. Of course one can always copy them in the C code using Rf_duplicate.

6.3   Writing C code to interface with R

The command R CMD SHLIB can be used to create a shared library suitable
for loading into R from a collection of C or FORTRAN program files. When
including source code in an R package, the recommended method is to place
all code in a subdirectory named src, and then the appropriate shared library
will be constructed when the package is installed.

6.3.1   Registering routines

One of the most important new developments in R is the ability to register
foreign routines. Registration provides a mechanism to help ensure that the
correct code is evaluated, and it also provides mechanisms for specifying the
type, but not currently the lengths, of different arguments. When the mechanism is used, only registered routines can be invoked and only those in the
dynamic library supplied with the package. Thus name collisions between different packages are avoided, and routines that are defined, but not registered,
cannot be inadvertently used. The **stats** package, supplied with R, provides
an exemplar where all routines are registered. Various Bioconductor packages
also make use of the registration mechanism.

The registration mechanism is relatively straightforward. The developer
must provide the appropriate C code for registration; then when the dynamic
library is loaded, that code will be invoked. This can be achieved by making
use of the fact that R will search for a C function named R_init_*mylib* and
invoke it when the shared library is loaded; see Section 6.5 for more details.
These are sometimes called initialization routines.

Routines are registered through a call to the C function R_registerRoutines.
This is typically done when the DLL is first loaded by placing a call to
R_registerRoutines within the initialization routine for the shared library.
R_registerRoutines takes five arguments. The first is the DllInfo object passed
by R to the initialization routine. This structure is used to store all the information about the different C and FORTRAN routines in the shared library
as well as other information about the shared library. You should not make
changes to the entries in this structure directly, but rather use the interface
routines provided. The remaining four arguments to R_registerRoutines are
arrays describing the routines for each of the four different interfaces: .C,
.Call, .Fortran and .External. Each argument is a NULL-terminated array
of the element types; see Table 6.2. For each one of the four interfaces that
your package or shared library uses, you will need to create a structure that
contains the appropriate information. That is, functions that are accessed by
.C will be described in a R_CMethodDef structure, while those that are accessed
by .External will be described in a R_ExternalMethodDef structure. Both the
R_CMethodDef structure and the R_FortranMethodDef structures have additional

Interface	Type
.C	R_CMethodDef
.Call	R_CallMethodDef
.Fortran	R_FortranMethodDef
.External	R_ExternalMethodDef

Table 6.2: Registration structures.

fields; one of which allows the developer to specify the types of the arguments. If the types are specified, then they will be checked whenever the function is invoked and an error will be signaled if one of the arguments has the wrong type.

For our example we will use one of the routines provided in the **stats** package. The function `cor.test` provides a number of different tests for association, among them Kendall's tau. The test itself is computed in C and is accessed via the .C interface. The code below first shows an R implementation, then the different pieces needed for registering routines. Much of the code has been omitted and readers are referred to the file `init.c` in the **stats** package. We do note that the C function takes three arguments: an integer, a real and an integer. These types are stored in an `R_NativePrimitiveArgType` array, where the values are SEXPTYPES. The appropriate SEXPTYPE can be determined from the entries in Table 6.1.

The call to `R_registerRoutines` registers routines accessed through the .C interface, the .Fortran interface and the .Call interface. The call to `R_useDynamicSymbols` indicates that if the correct C entry point is not found in the shared library, then an error should be signaled. Currently, the default behavior in R is to search all other loaded shared libraries for the symbol, which is fairly dangerous behavior. If you have registered all routines in your library, then you should set this to FALSE as is done in the **stats** package.

The code shown in Program 6.1 is based on code in the **stats** package and shows an R implementation of a function that computes distribution function for Kendall's tau.

In Program 6.2 we show an edited version of the relevant C routines. For the complete versions of this code, readers should obtain a source version of the stats package and investigate it.

6.3.2 Dealing with special values

In this section we address some of the solutions for dealing with special values such as missing values, infinity and NaNs. We discuss three separate topics here: first we discuss the special values that correspond to missing values and non-finite values, then we discuss passing out single precision values to C and lastly we discuss passing out matrices and arrays.

```
##R Code
  pkendall = function(q, n) {
     .C("pkendall",
        length(q),
        p = as.double(q),
        as.integer(n),
        PACKAGE = "stats")$p
}
```

Program 6.1: R code for pkendall.

```
/* define argument types */
static R_NativePrimitiveArgType pkendall_t[3] = {INTSXP, REALSXP,
    INTSXP};

/*list the name, C entry point, number of arguments
and their types */
static const R_CMethodDef CEntries[]  = {
 ...
 {"pkendall", (DL_FUNC)  &pkendall, 3, pkendall_t},
 ...
}

/* do the registration */
void R_init_stats(DllInfo *dll)
{
    R_registerRoutines(dll, CEntries, CallEntries, FortEntries,
        NULL);
    R_useDynamicSymbols(dll, FALSE);
}
```

Program 6.2: Edited C code relevant to the pkendall example.

We previously mentioned that for calls through .C and .Fortran, the user can set a flag, at the R level, that will prevent vectors containing missing values, or infinite values from being passed down. When writing C it is often helpful to deal with special values such as missing values or infinite values. R represents these special values differently, depending on the type of the vector, so users must both test and set values according to the type of the vector being processed.

For vectors of mode double, there are a number of macros that can be used: ISNA tests for NA, ISNAN tests for NaN or NA, and R_FINITE tests for NA and all the special values. Otherwise, direct comparison to the constants R_NaN, R_PosInf, R_NegInf and R_NaReal can be used. To test for integer missing values, compare to NA_INTEGER, for logicals compare to NA_LOGICAL and for character values compare to NA_STRING. These values and NA_REAL can be used to set elements of R vectors to be missing.

6.3.3 Single precision

All real values stored in R are stored in double-precision (the C data type is double), but in some cases it is desirable to use single precision at the C level, such as when a legacy implementation is being used. There are two R-level functions that can be used to indicate that the vector should be coerced to single precision when it is passed to C or FORTRAN. The first of these is as.single, and it attaches an attribute to the vector named Csingle with the value TRUE. Otherwise, empty vectors can be generated by a call to single, which produces a vector of zeros with the Csingle attribute set to TRUE. When a vector is passed out to either FORTRAN or C through the .Fortran or .C interface, it is checked for the existence of a Csingle attribute, and if that attribute exists and is set to TRUE, then the vector is coerced to single precision before the call is made.

6.3.4 Matrices and arrays

Neither FORTRAN nor C has an inherent notion of the matrix data structure, but rather the artifice of such a data structure can be maintained by the use of appropriate indexing into a one-dimensional array. This is precisely what R does for its internal data structures. However, C and FORTRAN differ in their notion of how the values in the vector are to be extracted. C uses what is called row major order, while FORTRAN uses column major order. The S language uses column major order, so when a matrix or an array is passed out through either .C or .Fortran, the function being called gets a vector and it must decide how to extract and manipulate the contents. Since these vectors are not self-describing, you will also need to pass out information about the number of rows and the number of columns, or more generally the extents of all dimensions. From these quantities the lengths of the vector can be deduced.

In the code below we provide a simple demonstration of the relationship between indices in R, and those needed to manipulate the same data in C. We first load the **RBioinf** package, since it has a number of simple functions predefined for just this purpose. We make use of the function simplePVect, which takes as input any vector, matrix or array of numeric values and passes that array out to C, where it is printed, in order from the first value stored to the last, regardless of the dimensioning information.

```
> library("RBioinf")
> x = matrix(1:6, nc = 2)
> x

     [,1] [,2]
[1,]    1    4
[2,]    2    5
[3,]    3    6

> simplePVect(x)
```

Wherein we see that the values are stored column by column. In printing the values from C, we retain C's use of zero-based subscripting, as you will need to deal with this in any computations you make. Thus, to find the value in the C vector that corresponds to the entry x[i,j], you will use the formula index = (i-1) + (j-1) * nrow.

Exercise 6.1
What is the correct formula for finding the entry x[i,j,k] if x is a three-dimensional array? Can you find the general formula for arrays of any dimensions?

Exercise 6.2
Write a C-level routine that computes row sums of a matrix. This is essentially a reimplementation of the rowSums function, so you can use it or apply to test your results. See the psuedo-code below for some help.

We provide some pseudo-code for the case where the matrix has **nrow** rows and **ncol** columns.

```
for(i=0; i<nrow; i++)
  for(j=0; j<ncol; j++)
    ans[i] += inmat[i+j*nrow];
```

So the values in the vector **ans** are the sums across all values in the corresponding rows. There are **ncol** values in each row, and those in row one of the matrix are found in positions $0, nrow, 2nrow, \ldots, (ncol - 1)(nrow)$. On the other hand, the elements of the first column (there are `nrow` of them) are found as the first `nrow` elements of the vector `inmat`. For higher dimensional arrays, it is always the left-most index that moves fastest, in the sense that `x[1,1,1]` is the first element in the array and `x[2,1,1]` is the second (provided of course the first dimension is larger than 1).

6.3.5 Allowing interrupts

One of the many things that occurs in R, by default, when calling to foreign functions is the turning off of all signal handling. For a very detailed discussion of signal handling under Unix, see Chapter 10 of Stevens and Rago (2005). Basically, by turning off the signal handling, it is no longer possible for a user to interrupt the running program. If your computations are likely to be long and involved, you should consider making use of the provided mechanisms to allow R to check for interrupts.

In C you must include the header file `R_ext/Utils.h`, and the name of the C routine to call is `R_CheckUserInterrupt(void)`. Placing calls to this function within your code will pass control back to R, where signals, such as those induced by control-C, are checked. You should be aware that this function may not return. If a signal has been raised, then typically R's error handling system is invoked and control is returned to the top-level prompt.

In FORTRAN, developers should use the subroutine `rchkusr()`.

Exercise 6.3
*Implement and test checking for user interrupts in one of the C routines provided in the package **RBioinf**.*

6.3.6 Error handling

One of the important, although often underdeveloped, areas of software engineering is providing informative error messages to the user. R has a very sophisticated error handling mechanism, although it is not well documented. We provide details of the functions available at the R-level in Section 2.11 and many of those R-level routines have corresponding C-level interfaces. Users can call either `error` or `warning`, with the same call sequences as to the C function `printf`.

6.3.7 R internals

R itself is primarily written in C. The overall design is reasonably simple, and the initial implementation began with a small Scheme interpreter, along the lines described in Kamin (1990). On top of that, a number of different data structures were layered, and then other functionality was gradually added. Essentially all language objects, data and virtually every object that is manipulated by the R interpreter are stored in a flexible structure, which we will refer to as a SEXP. A number of these are detailed in the fourth column of Table 6.1. Understanding the R internals or writing your own code to interact with the R internals requires at least a basic understanding of these different data types; see Section 6.3.7 for more details on R internals.

R manages its own memory, so creation of new instances of the different data structures involves calling specific internal functions to carry out the actual allocation. The design also requires that any routine that requests new storage be responsible for ensuring that the memory is marked as being in use. This is generally done by either assigning the new object into an object that is known to be marked or by explicitly using a macro, named `PROTECT`, to protect memory. Before exiting the routine, all calls to `PROTECT` must have a corresponding `UNPROTECT`. More details on the R internals are given in Section 6.3.7.

Some basic description of R internals is important, and here we provide a fairly cursory description of them. R has essentially one type of internal object, which is defined by the `SEXPREC` typedef. This structure is quite flexible and can represent almost all language structures and data structures. Most programming is done with `SEXP`s, which are pointers to `SEXPREC`s. The acronym `SEXP` stands for *S expression*, or *Symbolic expression*, and has its roots in Lisp (Steele, 1990) and Scheme (Abelson and Sussman, 1996) programming. The `SEXPREC` type provides a place to attach attributes as well as a number of other locations for linking together `SEXPREC`s to create some of the more complex data structures used in R. A partial list of the types available is given in Table 6.3.

The function `allocVector` should be used to allocate a vector of the desired type and length. Attributes can be added to these vectors to create different specialized types, e.g., adding a `dim` attribute modifies the vector to be an array. Generic vectors, `VECSXP`s, and strings, `STRSXP`s, are a little bit different than the other vectors. Essentially these two objects are collections of `SEXPREC`s of the requested length. The elements of the generic vector are then, themselves `SEXPREC`s. For `STRSXP`s, only `CHARSXP`s can be placed into the vector. `CHARSXP`s are special and should never be exposed directly at the R level; they should only ever be used as elements of `STRSXP`s.

Setting and getting values from `REALSXP`s, `INTSXP`s and `LGLSXP`s can be done by using the `REAL` macro for the first and the `INTEGER` macro for the second two. A small code segment demonstrating the use is given in Program 6.3. We demonstrate the use of three different types of simple vectors and the two more complicated types: generic vectors and strings. The only slightly tricky point is the use of pointers to reference the actual storage and their assignment outside of the first `for` loop. The reason you should consider this approach is that the macros `REAL` and `INTEGER` are real function calls, in user defined code, and hence incur a cost, which good optimizing compilers may be able to identify and remove.

R has a sophisticated garbage collection scheme and users should avoid, where possible, direct calls to `malloc` and other C-specific memory allocation

```
int *ivp, *lvp;
double *rvp;

PROTECT( iv = allocVector(INTSXP, 10) );
PROTECT( rv = allocVector(REALSXP, 10) );
PROTECT( lv = allocVector(LGLSXP, 10) );

ivp = INTEGER(iv); rvp = REAL(rv); lvp = INTEGER(lv);
for(i=0; i<10; i++ ) {
  ivp[i] = i;
  rvp[i] = i+10.0;
  lvp[i] = i % 1;
}

PROTECT( ml = allocVector(VECSXP, 3) );
SET_VECTOR_ELT(ml, 0, iv);
SET_VECTOR_ELT(ml, 1, rv);
SET_VECTOR_ELT(ml, 2, lv);

UNPROTECT(4);
PROTECT(ml);

PROTECT(sv = allocVector(STRSXP, 2) );
SET_STRING_ELT(sv, 0, mkChar("abc"));
SET_STRING_ELT(sv, 1, mkChar("def"));
```

Program 6.3: Pseudo-code showing how to access various internal R storage representations.

Name	Type
REALSXP	numeric with storage mode double
INTSXP	integer
CPLXSXP	complex
LGLSXP	logical
STRSXP	vector of character
VECSXP	list (generic vector)
LISTSXP	*dotted-pair* list
DOTSXP	a ... object
NILSXP	NULL (`R_NilValue`)
SYMSXP	a name or symbol
CLOSXP	function or function closure
ENVSXP	environment

Table 6.3: Some of the more widely used `SEXPREC` types.

mechanisms, as the memory obtained in that manner is of limited usefulness, and failure to explicitly free it can result in memory leaks.

R maintains two separate types of storage: one for the language structures (`SEXPREC`s) and the other for vector storage (primarily for storing data). When a request for memory is made, R first checks to see if there is allocated but unused memory available. If there is none, then a garbage collection is performed. Basically, memory that is in use is determined by tracing all symbols and also by examining values on a few special lists. One of those lists is the *protection* stack, and we shall describe its use shortly. After a garbage collection, any unused memory is identified as being available. If no memory is identified as being available, then new memory is requested from the system. If this fails, then R reports that more memory has been requested than is available and halts the current computation.

Because R manages memory, any author of C routines who wants to make use of R data structures must exercise discipline in their programming to ensure that all memory in use is clearly identified, and that when memory is no longer needed, it is made available back to the system. Carrying out these tasks is relatively simple and requires the use of two macros: PROTECT and UNPROTECT. A call to PROTECT with an SEXP as an argument places the object pointed to onto the protection stack. Note that the SEXPREC is protected, not the SEXP, so some care must be used to ensure that the pointer is not reused. Calls to UNPROTECT take an integer argument, and they pop the specified number of pointers off of the protection stack.

Protection is only needed if there are calls that could cause a garbage collection event. It is only when this happens that memory could be released or identified as free and reused for some other purpose. Thus, if there is only your own C code being used, there is no need to use the protect mechanism.

But, if you make any calls to R internal functions, and most often you will, then you must use the protect mechanism since you will not know whether those functions can trigger a garbage collection event.

Any function that has been invoked by either .External or .Call will have all of its arguments protected already. You do not need to protect them, and as mentioned above, they were not duplicated and should be treated as read-only values. If an object is protected, then so are all of its sub-components, including attributes, list or string elements and so on. For calls to .C and .Fortran, there is no need to worry about protecting or unprotecting. All arguments are protected prior to the call; and in most cases, the external routine will not have received the SEXP, but rather a pointer to some specific memory location, so that protection is not possible.

6.3.8 S4 OOP in C

OOP is an important programming paradigm and in R there are a number of macros that can be used to create and manipulate S4 classes and methods; see Chapter 3 for more details on S4. The macros are MAKE_CLASS, NEW_OBJECT, SET_SLOT and GET_SLOT.

The MAKE_CLASS macro retrieves the class definition for the named class. The class must be defined; this call merely retrieves that definition, which can then be used as input in the call to NEW_OBJECT. NEW_OBJECT creates an instance of the class supplied. The macros SET_SLOT and GET_SLOT can be used to access or modify the contents of the named slot.

An example of the use of most of these macros, from the **Biostrings** package, is given in Program 6.4.

```
SEXP mkBString(const char *class, SEXP data, int offset,
               int length)
{
        SEXP class_def, ans;

        class_def = MAKE_CLASS(class);
        PROTECT(ans = NEW_OBJECT(class_def));
        SET_SLOT(ans, mkChar("data"), data);
        SET_SLOT(ans, mkChar("offset"), ScalarInteger(offset));
        SET_SLOT(ans, mkChar("length"), ScalarInteger(length));
        UNPROTECT(1);
        return ans;
}
```

Program 6.4: An example from the **Biostrings** package that uses several of the S4 macros in C code.

6.3.9   Calling R from C

There are many situations where it will be convenient to be able to evaluate an R function from inside computations being carried out in C. We discuss this topic in some detail in Section 6.6.2.

6.4   Using the R API

The R application programming interface (API) is defined by the R Extensions manual and the information provided in the header file, `Rinternals.h`, provided with every R installation.

We now describe some ways in which you can simplify the task of writing C code. Perhaps the most important observation is that R contains very many utility routines that are widely tested and efficiently implemented. In most cases you should make use of these routines rather than reimplement them. Some additional reasons for using these routines are simplicity, less code development, no need to build rigorous testing paradigms, and by using routines in the R API you gain consistency because your answers from the C code will match answers obtained from prototype implementations in R.

Many of the details for making use of the R API are documented in the R Extensions manual. In this section we discuss some simple examples that are provided in the **RBioinf** package. These examples describe some interactions with the R API. Two of the most commonly needed tools are sorting and the generation of random numbers. There are two functions, `simpleSort` and `simpleRand`, provided.

The R API provides a variety of tools for sorting, sorting with indices and partial sorting. For random number generation, there are tools for generating Uniform, Normal and Exponential random variables. Users cannot directly access some of the RNG state information except through calls to R functions, and so our example here demonstrates calling back in to R.

6.4.1   Header files

Header files are also often referred to as *include* files since they are included in other C program files using the `include` directive. All R header files are in the `src/include` subdirectory of the source tree. Some of these header files are intended to be private to R, while others are intended to be public and used by developers who want to make use of R's internal data structures. When R is built, some of these header files are moved to `R_HOME/include`. The two main files are `R.h` and `Rinternals.h`. The first is needed when writing code to use the `.c` interface and the second is needed if any of R's internal data structures (e.g., SEXPs) will be manipulated. A number of other *include* files can be found in the `R_ext` subdirectory. You will need to `include` these if you want to make use of the corresponding functionality in R. For example, if the C registration mechanism, discussed below, is used, then `R_ext/Rdynload.h` must also be included.

6.4.2 Sorting

The function `simpleSort`, in Program 6.5, takes as input a vector of real-valued numbers, say `x`, and returns the index vector `ind` such that `x[ind]` is sorted from smallest to largest.

The C code is in the file `src/rand.c` of the source package for **RBioinf**. It is repeated below. The first computation is to check that the inputs are of the correct type. It is essential that these checks be carried out in C as it is very difficult to ensure that the function was invoked with the correct arguments. As the C registration mechanism becomes richer and more widely used, such checking will become less important. Next, the storage for the return value, `ans`, is allocated and initialized. When `ans` was initialized, we made sure to use one based indexing since that is what is used in R. Next we duplicated the input argument, since the sorting is destructive, and the input to `simpleSort` is not copied, we must make a copy. We do not protect the duplicated value, although we could, because there are no other memory allocations after it. The source code for `rsort_with_index` must be checked to be sure of that, and it may be better to be defensive and protect all duplicated values.

Exercise 6.4
Modify the C code in simple sort so that other types of input values can be used. Note that there is no easy way to directly access routines for sorting character vectors. Can you explain why that might be the case?

Exercise 6.5
*Create a new function, for addition into the **RBioinf** package, that does partial sorting. Define a suitable interface (possibly with a simple use-case scenario) and provide R and C implementations together with a manual page.*

6.4.3 Random numbers

Many of the pseudo-random number generating tools from R are directly available at the C level. Most do not require the use of R's internal data structures (e.g., `SEXP`) and can be accessed directly from C. The different functions available are well documented in the R Extensions manual and include simple functions for generating Normal, Uniform and Exponential random variates. Direct access to most of the specialized functions for different distribution functions (e.g., `dnorm`, `pnorm`, `rnorm` and `qnorm`) is also available.

Technically, all random number generators in R are pseudo-random number generators, but we will drop the prefix *pseudo* in our discussion and remind

```
SEXP simpleSort(SEXP inVec)
{
    SEXP ans, tmp;
    int i;

    if( !isReal(inVec) )
        error("expected a double vector");

    /* allocate the answer */
    PROTECT(ans = allocVector(INTSXP, length(inVec)));

    for(i=0; i<=length(inVec); i++)
        INTEGER(ans)[i] = i + 1; /* R uses one-based indexing */

    /* we need a copy since the sorting is destructive
       and inVec is not a copy */
    tmp = Rf_duplicate(inVec);

    rsort_with_index(REAL(tmp), INTEGER(ans), length(inVec));
    UNPROTECT(1);
    return(ans);
}
```

Program 6.5: simpleSort.

the reader that it is assumed. There are a number of different random number generators available, and which is deemed the best can change. If simulations are an important part of your application, some care should be taken to select an appropriate random number generator, possibly to allow the user to make his own choice, and to provide tools to manage the seed.

Pseudo-random number generators depend on a seed, and in R the seed is stored in the user's workspace as the variable `.Random.seed`. When R is started, there is no seed, and by default a random value is selected when any R function is used to generate random variates. Pseudo-random number generators are guaranteed to provide the same values if the generator is started with the same set of seeds. Users should be careful to manage seeds, and some tools are provided to make this possible. At the R level, the functions `set.seed` and `RNGkind` can be used to set a seed and to select the random number generator, respectively.

When calling any of the random number generating functions from C, or other external languages, it is the responsibility of the calling program to manage the seeds. All calls should be prefaced by a call to the C function `GetRNGstate` and followed by a call to `SetRNGstate`. The first of these calls sets up a random seed, if one does not exist. It will either create a new object or it will make use of the existing value to set up the random number generator. After random numbers have been generated, a call to `SetRNGstate` replaces the current value of `.Random.seed` with the appropriate value obtained from the random number generator. Without a call to `SetRNGstate`, the information about the state of the random number generator will be lost. The decision to have these functions manipulate a global variable, `.Random.seed`, is slightly unfortunate as it makes it somewhat more difficult to manage several different random number streams simultaneously.

In the **RBioinf** package there is an example of accessing some of the random number generating features from C. The example code shows how to access one of the random number generators and includes an example of calling back into R to determine what random number generator is in use. As noted above, the method used is perhaps better described as invoking a call to R's evaluator through the C function `eval`.

```
/*create the language structure needed to call R */
PROTECT(Rc = lang1(install("RNGkind")));
PROTECT(tmp = eval(Rc, R_GlobalEnv));
```

Program 6.6: Code showing how to determine which random number generator is in use.

In the code segment shown in Program 6.6, a language structure is allocated, and the symbol for the function we intend to call is made the first element of the `LANGSXP`. Since the default values of all arguments for the R function `RNGkind` will be used, only the name of the function is used. The returned value, `Rc`, is `PROTECT`ed because the call to `eval` will require some allocation and hence could trigger a garbage collection event. To ensure that the memory addressed by `Rc` is not garbage collected, it is protected. The return value from `eval` is also protected, since it will have been newly allocated, and it is the responsibility of the calling function to protect return values. Interested readers should examine the source code for the entire C function `simpleRand` for further details.

Exercise 6.6
Modify the C and R source code, as well as the manual pages, for `simpleRand` *so that the user can specify one of the three basic random number generators at the C level.*

Harder: Modify the C source code so that the user can set the type of random number generator through a call to `eval` *for* `RNGkind`.

Exercise 6.7
Modify the C and R sources for `simpleRand` *so that it manages its own seed. Care must be taken to save the global seed, replace it with the supplied one, generate random numbers, save the seed, restore the global seed and return a data structure with the seed in it.*

6.5   Loading libraries

Shared libraries can be loaded into R using the `dyn.load` function. In many cases the shared library is part of a package and then `library.dynam` is preferred, and it calls `dyn.load` after carrying out some other computations appropriate for loading R packages. When the shared library is no longer required, it can be detached using `dyn.unload`, and for packages the equivalent function is `library.dynam.unload`. More details on building packages with C or FORTRAN code are provided in Chapter 7.

When `dyn.load` is invoked, say as `dyn.load("xxx.so")`, then a shared library is loaded into R. Once that is completed, R looks for a function, or entry point, with prefix `R_init_` and with suffix the name of the object loaded, without the system-specific suffix for shared libraries, and invokes it. Hence in our example, R would look for a C function named `R_init_xxx`, and if it exists, it will be run. This provides a mechanism that allows the developer of the C code to provide initialization routines and to perform any computations needed before the code is ready to use.

There is a similar mechanism that can be used to carry out computations when a shared library is unloaded or detached. R looks for a C function named

R_unload_ with suffix the name of the shared library. So, in our example, R would look for a C routine named R_unload_xxx, and if it is found, it will be invoked.

Both of the routines, R_init_ and R_unload_, are called with a single argument, which is a pointer to a DllInfo structure. This structure contains relevant information for the shared library, and the load and unload code can make use of any information contained in it.

Exercise 6.8
*In the Exercises directory of the **RBioinf** package there is a template file named xxx.c. Use this file, together with R CMD SHLIB, to build a shared library, and load it into R. What happens? Rename the file to be foo.c and repeat. What happens, and why?*

6.5.1 Inspecting DLLs

R provides tools that let the user find out which DLLs are loaded, and for any of those to obtain the set of registered routines. The set of loaded DLLs is found by calling getLoadedDLLs.

```
> getLoadedDLLs()


                                                            Filen
                                                             ame
base                                                            b
ase
methods          /Users/robert/R/R27/library/methods/libs/methods
.so
grDevices /Users/robert/R/R27/library/grDevices/libs/grDevices
.so
stats              /Users/robert/R/R27/library/stats/libs/stats
.so
cluster          /Users/robert/R/R27/library/cluster/libs/cluster
.so
tools              /Users/robert/R/R27/library/tools/libs/tools
.so
graph              /Users/robert/R/R27/library/graph/libs/graph
.so
RBioinf          /Users/robert/R/R27/library/RBioinf/libs/RBioinf
.so
Biobase          /Users/robert/R/R27/library/Biobase/libs/Biobase
.so
              Dynamic.Lookup
```

base	FALSE
methods	FALSE
grDevices	FALSE
stats	FALSE
cluster	FALSE
tools	FALSE
graph	TRUE
RBioinf	TRUE
Biobase	TRUE

Once the DLL is loaded, then `getDLLRegisteredRoutines` can be used to obtain information on the registered routines that are contained in the DLL. This can be quite helpful for debugging purposes.

6.6   Advanced topics

In this section we discuss a few subjects that are of a more advanced nature but which are often useful when writing C, or other code, to interface with R. The topics include allowing users to interrupt long-running processes, a discussion of external references and a more detailed discussion of constructing R expressions in C and evaluating them.

6.6.1   External references and finalizers

This section is primarily based on the notes *Simple References with Finalization* provided by L. Tierney and parts of both the semantics and the implementation have not been completely worked out yet. However, the system is useful and provides an important tool when working with large data, or other objects where it is useful to have reference semantics. Reference semantics can also be achieved using environments in a structured way.

An external reference is an R object, of type `EXTPTRSXP`, that holds a reference, or pointer, to an external resource. There is no R-level interface; all interactions are at the C level. At the R level, these objects have type `externalptr`. They, like environments, are not copied, so there is only one version of them. It should be noted that this affects the usefulness of attributes on such objects, as they too are not copied. Thus, if you want to create an R object that contains one of these and on which attributes, etc., can be usefully attached, then they should be either enclosed in a list or placed into an S4 style object.

At the C level, an external pointer is constructed by calling `R_MakeExternalPtr` with three arguments: the pointer value, a tag SEXP, and a value (an SEXP) to be protected. The `prot` argument is protected from garbage collection for the life of the external pointer, and as we shall see below, this provides a useful way to create external pointers using memory allocated by R. The tag can be used to attach other information such as type, length, etc. to the pointer. The values of each of these three elements should be accessed using the accessor functions defined in the code shown in Program 6.7.

The precise semantics of saving and restoring external references are not yet defined, so some care should be exercised when using them in settings that might involve quitting from R and starting again at a later time.

```
SEXP R_MakeExternalPtr(void *p, SEXP tag, SEXP prot);

void *R_ExternalPtrAddr(SEXP s);
SEXP R_ExternalPtrTag(SEXP s);
SEXP R_ExternalPtrProtected(SEXP s);

SEXP R_AllocatePtr(size_t nmemb, size_t size, SEXP tag)
{
    SEXP data, val;
    int bytes;
    if (INT_MAX / size < nmemb)
        error("allocation request is too large");
    bytes = nmemb * size;
    PROTECT(data = allocString(bytes));
    memset(CHAR(data), 0, bytes);
    val = R_MakeExternalPtr(CHAR(data), tag, data);
    UNPROTECT(1);
    return val;
}

/* Finalizer Code */

typedef void (*R_CFinalizer_t)(SEXP);

void R_RegisterFinalizer(SEXP s, SEXP fun);
void R_RegisterCFinalizer(SEXP s, R_CFinalizer_t fun);
```

Program 6.7: Code snippets for external references and finalizers, largely repeated from Tierney (2002).

The C function `R_AllocatePtr` shows how to use this system to allocate memory using R, in this case the `allocString` function was used, and to create an external vector using that memory.

6.6.1.1 Finalizers

A finalizer is a piece of code, generally a function, that is run when an object has ended its useful life and is about to be garbage collected. In R, finalizers can be registered for either environments or for external references. When the corresponding object is identified as no longer being in use, the finalizer will be run. However, there is no guarantee as to the order in which finalizers are run. It is also important to realize that finalizers may not be run when an R session is ended. Explicit invocation of a garbage collection, via a call to gc, will cause all identified finalizers to run.

Finalizers are useful for ensuring that some cleaning up is done. For example, should the memory allocated in an external pointer be allocated using a call to malloc, then it will need to be explicitly freed and the finalizer can be set up to do that.

6.6.2 Evaluating R expressions from C

It is sometimes helpful to be able to evaluate R expressions from within C, or other foreign language, interfaces. This process is sometimes referred to as calling R, but it seems to be more precise to describe it as evaluating R expressions. This is a fairly complex task and definitely not one for beginners. While there is an interface provided, via call_R, this is seldom used because direct access and manipulation is generally much easier and more effective. We take that same view and describe how to carry out such direct manipulations.

Some of the steps that you will need to carry out include:

- creating R expressions

- defining symbols within an R environment

- evaluating an R expression within an R environment

6.6.2.1 Creating R variables in environments

This is a reasonably easy task. In R, symbols, such as x, are linked to values via their bindings in environments. To look up the value of a symbol, you often must first translate from a character representation to the actual internal R symbol.

Program 6.8 defines a small C function that takes as input an environment, rho, and looks for a value bound to the symbol x in that environment. You should recall, from Chapter 2, that environments have parents; and if a value for x is not found in rho, then the parent of rho will be searched and so on. If you do not want this behavior, then you can use the C function findVarInFrame, which only looks in the environment provided.

The second part of the code in the example accesses the symbol y, creates an integer vector of length 1, assigns the value 10 into that vector, and then assigns that as the value of y in the environment rho. Since there is some chance that new memory will need to be allocated in the call to defineVar, we must protect the allocated memory (assigned to Rv) from garbage collection until it is safely stored in rho. We do assume that rho itself will not be garbage collected.

```
SEXP getX(SEXP rho)
{
SEXP Rsym, Rval, Rv;

Rsym = install("x");
Rval = findVar(Rsym, rho);

Rsym = install("y");
PROTECT(Rv = allocVector(INTSXP, 1));
INTEGER(Rv)[0] = 10;
defineVar(Rsym, Rv, rho);
UNPROTECT(1);

return(Rval);
}
```

Program 6.8: Pseudo-code for finding symbols.

6.6.2.2 Evaluation

We now consider the process of evaluating an expression in a foreign language. We will hold off on function evaluation for a little while.

Program 6.9 defines a very simple function, evalExpr. We assume that an expression is passed to this function as well as an environment in which to evaluate it. Suppose we have compiled and made this C function available in R, and then invoke it with the following sets of commands.

```
> e1 = new.env()
> e1$x = 1:10
> .Call("evalExpr", quote(sum(x)), e1)
```

Exercise 6.9

What will the answer be? Can you explain why we used quote? *In fact, this function is available through the* **RBioinf** *package. Try a more complex evaluation, say adding together values for* x *and* y.

```
SEXP evalExpr(SEXP expr, SEXP rho)
{
  return(eval(expr, rho))
}
```

Program 6.9: Pseudo-code showing how to evaluate an expression in C.

6.6.2.3 Constructing function calls

In R, there are essentially two types of lists: the one used most often at the R level is essentially a generic vector, while the other is a Lisp-style list that is sometimes referred to as a `pairlist`. While seldom used at the R level, most of R's internals are written using pairlists. And in particular, this is the internal representation used for functions, expressions, calls and other objects. To understand how to construct and manipulate these at the C level, we first provide a very brief description of how they work.

Each element of a pairlist consists of three parts: the CAR, the CDR and the TAG. By convention, the CAR points to the value stored in that element, the CDR points to the next element in the list (or `R_NilValue` if there is no next element), and the TAG contains the name associated with the value, if there is one. There are a variety of macros that can be used to access these elements. The most widely used are CAR, CDR and TAG, but others such as CADR (the CAR of the CDR) and CDDR (the CDR of the CDR) are also defined and should be used.

R language elements are essentially `pairlists` but they have a different type, typically `LANGSXP` objects. To construct a language element, you can either call one of the builtin `langX` functions (X can be any integer between 1 and 4) or create a pairlist of the desired length using `allocList` and then set its type to be `LANGSXP` (e.g., `SET_TYPEOF(s, LANGSXP)`).

A function call requires a LANGSXP of length equal to one plus the number of actual arguments. The symbol that identifies the function goes in the first element and actual arguments fill the remaining elements of the pairlist. In Program 6.10 we show some pseudo-code for setting up a function call with three arguments, two of them named. When we first allocate the pairlist, we protect it from garbage collection, since we will be making some future allocations and hence could trigger a garbage collection.

We make use of an additional pointer, t, as if we must keep track of where the top of the pairlist is, the only access to that is through `myfun`. The values put into the TAG components are symbols, and hence we must use `install` to get those.

```
SEXP myfun, t;

PROTECT(t = myfun = allocList(4));
SET_TYPEOF(myfun, LANGSXP);
CAR(myfun) = install("mean");
t = CDR(t); CAR(t) = allocVector(REALSXP, 1);
REAL(CAR(t))[0] = 101; TAG(t) = install("v1");
t=CDR(t); CAR(t) = allocVector(REALSXP, 2);
REAL(CAR(t))[0] = 9; REAL(CAR(t))[1] = 12.5;
TAG(t) = install("v2"); t = CDR(t);
CAR(t) = allocVector(REALSXP, 1);
REAL(CAR(t))[0] = -2;

eval(myfun, rho);
```

Program 6.10: Pseudo-code for creating a function call.

Exercise 6.10
Implement the zero finding algorithm discussed in the R Extensions manual.

6.7 Other languages

Our treatment of this topic is brief and readers are referred to Chapter 12 of Chambers (2008) for a more extensive discussion. There are interfaces between R and Java, Perl, and Python that are somewhat widely used. But it should be noted that these interfaces rely on the R-C interface, and that they are typically implemented via a C language interface. For example, to communicate between R and Python, the **RSPython** package uses the function `.Python`, whose definition is shown below.

This uses the `.Call` interface, and shows that there is a C-level interface that is used.

Duncan Temple Lang is the author of several of foreign language interface packages, and his packages can be obtained from the Omegahat web site,

```
.Python =
#
# Invoke a Python function.
#
#
function(func, ..., .module = NULL, .convert = TRUE)
{
 isPythonInitialized(TRUE)

 args = list(...)
 .Call("RPy_callPython", as.character(func), args, names(args),
       as.character(.module), as.logical(.convert), FALSE)
}
```

`www.omegahat.org`. A comprehensive listing of all packages available through Omegahat can be obtained using the first command below. The same URL can be used in `download.packages`, `install.packages`, or `update.packages` to obtain copies of the most recent versions.

```
> available.packages(contrib.url("http://www.omegahat.org/R",
      type = "source"))
```

The three packages **RSJava**, **RSPerl** and **RSPython** are all bidirectional. That is, they provide an interface that either allows R to call into the other language, or code in the other language to call R.

The **RPy** was inspired by the **RSPython** package, but implements only the interface from Python to R and cannot be used from R to call Python code.

Simon Urbanek provides **JRI** and **rJava**. The former provides an interface from Java to R, while the latter provides an interface from R to Java. They seem to be somewhat easier to use and more robust than the **RSJava** package, although the latter provides a more comprehensive foreign language interface. In all cases, the major issues users report tend to be related to having the correct CLASSPATH set for Java and appropriate environment variables set for R.

Chapter 7

R Packages

An R package typically consists of a coherent collection of functions and data structures that are suitable for addressing a particular problem. Packages are easy to distribute and share with others, and in many ways writing a package is the best and most effective way to share your software and ideas with others. Learning to move away from collections of scripts, or functions, that get sourced to software organized in packages is very enabling. Writing your first package can be a challenge, but the second and third are simpler. There are many advantages to writing a package, including the fact that everyone you collaborate with can be using precisely the same code and all can contribute to its quality and usability.

In many cases, developers will either have access to existing code written in some other language, or they may find that they need to implement some parts of their software in C, or some other compilable language, for efficiency. This topic is covered in Chapter 6.

The many hundreds of contributed packages provide a great resource, and for many problems you will simply want to find, download, install and use packages written by others. There are currently three main repositories for R packages: CRAN, Bioconductor and Omegahat. CRAN contains the most R packages (over 1000) while Bioconductor and Omegahat are smaller. The number of packages is somewhat overwhelming and two related innovations, CRAN Task Views (the **ctv** package) and the **biocViews** packages, provide tools that can be used to provide overviews of packages suitable for different subject areas. Both approaches use restricted sets of terms to help organize packages. We will discuss in Section 7.2.1 some of the ways in which **biocViews** can be used and readers who visit the Bioconductor web site will see it in action, as the **biocViews** terms are used to organize the Bioconductor packages.

In this chapter we first review some of the available functionality for using packages and then discuss how to write your own packages. For most of the topics we will discuss, the R Extensions Manual (R Development Core Team, 2007c) is a more comprehensive reference.

7.1   Package basics

An R package is basically a collection of folders that contain the R code, the
help pages, data, other documentation such as vignettes, code written in other
languages such as C or FORTRAN, and a number of files that provide directives used to help install the package. Packages should run on all platforms
supported by R and there are a number of tools to help authors modify their
package to run on different platforms. All packages must have an R directory
and a man directory. If they provide data sets, then they have a data directory; and if they have source code for another language, then they will have an
src directory. Vignettes go in the inst/doc directory and this is required for
packages submitted to the Bioconductor project. Both the man directory and
the R directories can contain system-specific files in subdirectories named unix
and windows. All packages must contain a DESCRIPTION file, which contains
information on the package such as version number, author, maintainer, and
dependencies. Packages can have a NAMESPACE file that is used to restrict the
symbols that are exported, and to import functionality from other packages;
name spaces are discussed in more detail in Section 7.3.4.

It is common for packages to depend on other packages for some functionality. Dependencies can be specified in two different ways. First, there are two
fields in the DESCRIPTION file, Depends and Suggests, that can be used to list
other packages that provide important functionality. With the introduction of
name spaces came another way to indicate functional dependencies and that
is to use an import directive in the NAMESPACE file, and packages named here
should also be listed in the Imports field of the DESCRIPTION file. It is also
possible to implicitly import a package by using the double colon (::) operator in R code contained in the package, but this is discouraged as it makes
it difficult to programmatically detect the dependencies. Perhaps the major
difference between depending on a package and importing it is that for the
former, the package is attached to the search path, while for the latter it is
loaded but does not appear on the search path. We will make the distinction
clearer later in this chapter.

A recent innovation is support for a NEWS file that can either be located in
the top-level package folder or in the inst directory. This file should document
changes in the package so that users can easily find out what improvements
or modifications have been made. The format should be the same as that of
the NEWS file that comes with R.

7.1.1   The search path

There is a small set of default packages that are attached every time R is
started interactively. Currently, they consist of **methods**, **stats**, **graphics**,
grDevices, **utils**, **datasets** and **base**. When R is run in batch mode, not

all of these packages will be attached. You can find out what packages are currently attached to the search path by using the function `search`, and more detailed information can be obtained by using `sessionInfo`.

```
> search()

[1] ".GlobalEnv"          "package:stats"
[3] "package:graphics"    "package:grDevices"
[5] "package:utils"       "package:datasets"
[7] "package:methods"     "Autoloads"
[9] "package:base"
```

Earlier, in Chapter 2, we discussed the evaluation model used by R. We now need to distinguish between attaching a package to the search path and loading a package. The two terms used to be used interchangeably, but now we need to make a distinction. Packages are attached to the search path by direct calls to either `library` or `require`. When a package is attached, then all of its dependencies (as determined by the `Depends` field in its `DESCRIPTION` file) are also attached. Such packages are part of the evaluation environment and will be searched. But, another way to satisfy dependencies is to use an import statement in the `NAMESPACE` file. Such packages are said to be loaded, but are not attached. And the main distinction is that imported packages do not appear on the search path and are essentially available only to the package that imports them.

7.1.2 Package information

Many functions can generate information about packages that have already been installed on the user's system. A vector listing the base names of packages that are currently attached can be obtained using `.packages`. The return value of `.packages` is invisible, so we first assign it to a temporary variable, as in the example below.

```
> z = .packages()
> z

[1] "Biobase"    "tools"      "stats"      "graphics"
[5] "grDevices"  "utils"      "datasets"   "methods"
[9] "base"
```

Exercise 7.1
Compare the output of `.packages`, `search` and `sessionInfo`.

The path, or full location, of a package can be obtained using the `system.file` function. This can be particularly useful for finding documentation, or additional files that have been supplied with the package.

```
> system.file(package = "tools")

[1] "/Users/robert/R/R27/library/tools"
```

The contents of a package's DESCRIPTION file can be obtained using the function `packageDescription`.

```
> packageDescription("base")
```

```
Package: base
Version: 2.7.0
Priority: base
Title: The R Base Package
Author: R Development Core Team and contributors
      worldwide
Maintainer: R Core Team <R-core@r-project.org>
Description: Base R functions
License: GPL (>= 2)
Built: R 2.7.0; ; Mon Jun 9 10:37:52 PDT 2008; unix

-- File: /Users/robert/R/R27/library/base/Meta/package.rds
```

```
> packageDescription("base", fields = c("Package",
      "Maintainer"))
```

```
Package: base
Maintainer: R Core Team <R-core@r-project.org>

-- File: /Users/robert/R/R27/library/base/Meta/package.rds
-- Fields read: Package, Maintainer
```

The `fields` argument restricts the returned values to only those fields that are specified.

7.1.3 Data and demos

There are three functions that search all packages for specific types of information. The three types of information are data files, which can be searched for using the `data` function; demo files, which can be searched for using the `demo` function; and vignettes, which can be searched for using the R function `vignette`. We briefly describe the first two of these, as they are somewhat more well established, and fairly easy to describe, and spend more time on vignettes in the next section.

Data files are often supplied with different packages. They are typically used to demonstrate the use of different statistical methods. Prior to the development of the lazy loading mechanism, one had to access a data set through a call to `data` with the name of the data set wanted as an argument. If the package containing the data set of interest has set `LazyData` to `true` (Section 7.3.1), then the data set can be accessed by name; there is no need to invoke `data`. However, even in this case, there is sometimes a need to use `data` with the data set name as an argument. For example, if you have altered the data set and want to retrieve the original version, you must use `data`, since simply typing the name of the data set can only retrieve the altered version.

Demos are R scripts that can be run from inside of R. They typically demonstrate the use of some of the software in a package. Particularly nice examples are the demos in the **graphics** and **lattice** packages.

7.1.4 Vignettes

A vignette is a document that integrates code and text and describes how to perform a specific task. The help pages for a package should tell a user how to call specific functions that are included in the package, but they typically are a poor medium in which to express the use of the different package components in a coordinated way to carry out a particular analysis. Vignettes were developed as part of the Bioconductor Project primarily for this purpose. They have since been well established as part of the R system.

The function `Sweave` from the **utils** package is perhaps the most widely used tool for literate programming in R. The **relax** package provides a different interface, relying on Tcl/Tk to provide an easy-to-use interface. More recently, the **odfWeave** has been made available. It provides `Sweave` processing of Open Document Format files.

The **weaver** package provides additional tools and extensions that can ease some of the pain of authoring an `Sweave` document. In particular, it provides a mechanism that caches the computations for code chunks, so that on subsequent runs code chunks that have not changed do not need to be re-evaluated. Without the **weaver** package, every code chunk is evaluated every time the document is processed, even if only the text was altered.

In Gentleman and Temple Lang (2007) and Gentleman (2005), we extended the concept of an interactive document to the notion of a *compendium*. A compendium is essentially a collection of code, data and text that can be used to recreate a specific analysis in complete detail. One implementation of a compendium is as an R package, where there are structured ways to include code, data and a literate document, such as an `Sweave` document.

The `vignette` function has a `print` method that will open the PDF version of a vignette for reading. It also has an `edit` method that will extract the code chunks from the vignette and open them in an editor, thereby facilitating cut and paste operations.

In the **tools** package, there is a function `pkgVignettes`, which can be used to locate the vignettes and provides directory path details on all vignettes in a package.

```
> library("genefilter")
> pkgVignettes("genefilter")


$docs
[1] "/Users/robert/R/R27/library/genefilter/doc/howtogenefilte
r.Rnw"
[2] "/Users/robert/R/R27/library/genefilter/doc/howtogenefinde
r.Rnw"

$dir
[1] "/Users/robert/R/R27/library/genefilter/doc"

attr(,"class")
[1] "pkgVignettes"
```

Exercise 7.2
*How many vignettes are provided with the **Biobase** package? Open one of these in a PDF viewer, and a different one using the `edit` method.*

7.2   Package management

Collections of packages are stored in *libraries*. The most common location for a library folder is the default library that is installed when R is. Its location can easily be determined by calling `system.file` with no arguments, which will report the location of the **base** package. Other libraries can be created simply by creating a folder and instructing R to install packages in that location. If packages are being installed from the command line, then the `-library` flag can be used. If packages are being installed using `install.packages` the argument `lib` can be set appropriately.

Users of Bioconductor are strongly urged to use `biocLite`, as described on the Bioconductor web site. It greatly simplifies the interactions with packages and ensures that the packages you obtain are suitable for the version of R you have installed. But in many cases you will need to interact directly with the R package management system, and it is described next.

The package management system provides a series of functions to install, update and remove packages, as well as other tasks. These packages require that a repository (an online source for downloadable software) be specified. The option `repos` can be set to the appropriate URL or the repositories can be selected using the `setRepositories` function. On some platforms, selecting a repository and downloading packages from it can be done using the menu system, or `Tcl\Tk` widgets.

Once a repository has been specified, a variety of tools can be used for automatically updating, downloading and installing software from them.

download.packages downloads specified packages from the repository.

install.packages installs the specified packages from the repository.

update.packages checks to see if any installed packages are out of date with respect to the available packages in the repository, and updates if necessary.

remove.packages uninstalls the specified packages.

available.packages returns a matrix containing information about available packages in a repository.

old.packages compares the output of `available.packages` to that of `installed.packages` and reports those installed packages for which newer versions are available.

new.packages does the same comparison as `old.packages` but reports uninstalled packages that are available.

installed.packages lists packages installed on the user's system.

There are a number of functions in the **tools** package that can be used to determine whether package dependencies are met, or can be met from a specified repository. The easiest to use, for working with a single package, is `pkgDepends`. The function `package.dependencies` should be used for testing many packages.

```
> library("Biobase")
> pkgDepends("Biobase")
```

```
$Depends
[1] "methods" "tools"   "utils"

$Installed
[1] "methods" "tools"   "utils"

$Found
list()

$NotFound
character(0)

$R
[1] "R (>= 2.7.0)"

attr(,"class")
[1] "DependsList" "list"
```

A comprehensive and graphical overview of the package dependencies can be obtained using the **pkgDepTools** package. This package parses information from a CRAN-style package repository and uses that to build a dependency graph based on the Depends field, the Suggests field or both. Then tools in the **graph**, **RBGL** and **Rgraphviz** packages can be used to find paths through the graph, locate subgraphs, reverse the order of edges to find the packages that depend on a specified package and many other tasks.

7.2.1   biocViews

All contributors to the Bioconductor Project are asked to choose a set of terms from the set of terms that are currently available. These can be found under the Developer link at the Bioconductor web site. The terms are arranged in a hierarchy.

These are then included in the DESCRIPTION file. Below we show the relevant entry for the **limma** package, which is one of the longer ones.

```
biocViews: Microarray, OneChannel, TwoChannel, DataImport,
    QualityControl, Preprocessing, Statistics,
    DifferentialExpression, MultipleComparisons, TimeCourse
```

These specifications are then used when constructing the web pages used to find and download packages. An interested user can select topics and view only that subset of packages that has the corresponding **biocViews** term.

7.2.2 Managing libraries

In many situations it makes sense to maintain one or more libraries in addition to the standard library. One case is when there is a system level R that all users access, but all users are expected to maintain their own sets of add-on packages. The location of the default library can be obtained from the variable `.Library`, while the current library search path can be accessed, or modified, via the `.libPaths` function.

```
> .Library

[1] "/Users/robert/R/R27/library"

> .libPaths()

[1] "/Users/robert/R/R27/library"
```

The environment variable `R_LIBS` is used to initialize the library search path when R is started. The value should be a colon-separated list of directories. If the environment variable `R_LIBS_USER` is defined, then the directories listed there are added after those defined in `R_LIBS`. It is possible to define version-specific information so that different libraries are used for different versions of R, in `R_LIBS_USER`. Site-specific libraries should be specified by the environment variable `R_LIBS_SITE`, and that controls the value of the R variable `.Library.site`. Explicit details are given in the R Extensions manual.

7.3 Package authoring

Authoring a package requires that all of the different components of a package, many described above, be created and assembled appropriately. One easy way to start is to use the `package.skeleton` function, which can be used when creating a new package. This function creates a skeleton directory structure, which can then be filled in. A list of functions or data objects that you would like to include in the package can be specified and appropriate entries in the R and data directories will be made, as will stub documentation files. It will also create a `Read-and-delete-me` file in the directory that details further instructions. The R Extensions manual provides very detailed and explicit information on the requirements of package building. In this section we will concentrate on providing some general guidance and in describing the strategies we have used to create a large number of packages. Perhaps the easiest way to create a package is to examine an existing package and to modify it to suit your needs.

If you have published a paper describing your package, or have a particular way that you want to have your package cited, then you should use the functionality provided by the `citation` function. If you provide a `CITATION` file, then it will be accessed by the `citation` function.

7.3.1 The `DESCRIPTION` file

Every R package must contain a `DESCRIPTION` file. The format of the `DESCRIPTION` file is as an ASCII file with field names followed by a colon and followed by the field values. Continuation lines must begin with a space or a tab. The `Package`, `Version`, `License`, `Description`, `Title`, `Author`, and `Maintainer` fields are mandatory, all other fields are optional. Widely used optional fields are `Depends`, which lists packages that the package depends on, `Collate` which specifies the order in which to collate the files in the R subdirectory.

Packages listed in the `Depends` field are attached in the order in which they are listed in that field, and prior to attaching the package itself. This ensures that all dependencies are further down the search path than the package being attached. The `Imports` field should list all packages that will be imported, either explicitly via an imports directive in the `NAMESPACE` file, or implicitly via a call to the double-colon operator. The `Suggests` field should contain all packages that are needed for package checking, but are not needed for the package to be attached.

Lazy loading (Ripley, 2004) is a method that allows R to load either data or code, essentially on demand. Whether or not your package uses lazy loading for data is controlled by the `LazyData` field, while lazy loading of code is controlled by the `LazyLoad` field; use either `yes` or `true` to turn lazy loading on, and either `no` or `false` to ensure it is not used. If your package contains S4 classes or makes use of the **methods** package, then you should set `LazyLoad` to `yes` so that the effort of constructing classes and generic functions is done at install time and not every time the package is loaded or attached.

If the `LazyLoad` field in the `DESCRIPTION` file is set to `true`, then when the package is installed all code is parsed and a database consisting of two binary files, `filebase.rdb`, which contains the objects, and `filebase.rdx`, which contains an index, is created.

7.3.2 R code

All R code for a package goes into the R subdirectory. The organization there is entirely up to the author. We have found it useful to place all class definitions in one file and to place all generic function definitions in one file. This makes them easy to find, and it is then relatively easy to determine the class structure and capabilities of a package.

In some cases, some files will need to be loaded before others; classes need to be defined before their extensions or before any method that dispatches on them is defined. To control the order in which the files are collated, for loading into R, use the `Collate` directive in the `DESCRIPTION` file.

If there is code that is needed on a specific platform, it should be placed in an appropriately named subdirectory of the `R` directory. The possible names are `unix` and `windows`. There should be corresponding subdirectories of the `man` directory to hold the help pages for functions defined only for a specific platform, or where the argument list or some other features of the function behave in a platform-specific manner.

In addition, there are often operations that must occur at the time that the package is loaded into R, or when it is built. There are different functions that can be used, depending on whether or not the package has a name space. These are described in Section 7.4.

7.3.3 Documentation

The importance of good documentation cannot be overemphasized. It is unfortunate that this is often the part of software engineering that is overlooked, left to last, and seldom updated. Fortunately the R package building and checking tools do comparisons between code and documentation and find many errors and omissions.

We divide our discussion of documentation into two parts; one has to do with the documentation of functions and their specific capabilities while the other has to do with documenting how to make use of the set of functions that are provided in an R package. Function documentation should concentrate on describing what the inputs are and what outputs are generated by any function, while vignettes should concentrate on describing the sorts of tasks that the code in the package can perform. Vignettes should describe how the functions work together, possibly with code from other packages, to achieve particular analyses or computational objectives. It is reasonable for a package to have a number of vignettes if the code can be used for different purposes.

In R, the standard is to have one help page per function, data set, or important variable, although sometimes similar concepts will be discussed on a single help page. These help pages use a well-defined syntax that is similar to that of LATEX and is often referred to as `Rd` format, since that is the suffix that is used for R documentation files. The `Rd` format is exhaustively documented in the R Extensions manual. It is often useful to include at least one small data set with your package so that it can be used for the examples.

Once the R code has been written, a template help page is easily constructed using the `prompt` function. The help page files created by `prompt` require hand editing to tailor them to the capabilities and descriptions of the specific functions. The function `promptPackage` can be used to provide a template file of documentation for the package. Other specialized prompt functions include `promptClass` and `promptMethods` for documenting S4 classes and methods.

```
\name{channel}
\alias{channel}
\title{Create a new ExpressionSet instance by selecting
          a specific channel}
\description{
  This generic function extracts a specific element from an
  object, returning a instance of the ExpressionSet class.
}
\usage{
channel(object, name, ...)
}
\arguments{
  \item{object}{An S4 object, typically derived from class
    \code{\link{eSet}}}
  \item{name}{The name of the channel,
                a (length one) character vector.}
  \item{...}{Additional arguments.}
}
\value{
  An instance of class \code{\link{ExpressionSet}}.
}
\author{Biocore}
\examples{
obj <- new("NChannelSet",
          R=matrix(runif(100), 20, 5),
          G=matrix(runif(100), 20, 5))
##the G channel as an ExpressionSet
channel(obj, "G")
  }
\keyword{manip}
```

Program 7.1: The manual page for the channel function in the **Biobase** package.

The documentation for every function should include one or more realistic examples that show how to use it, on a variety of inputs. These examples perform two functions: first they allow the user to see how to call the function and, second, through the R CMD check procedure these examples are run and thus they also form part of the quality control procedures. These are very important and you should make some effort to choose examples that ensure that your code is performing as expected. If, at some later date, you make changes to the underlying code, a failure in R CMD check will alert you to potential problems for your users that should be remedied.

In Program 7.1 we show the help page for the channel function from the **Biobase** package. While many developers use file names that correspond to the name of the object being documented, the file name is irrelevant and you should feel free to name them as you like. Every help page must contain a \name field, and this name must be unique within a package, but it is used for internal purposes and also does not need to correspond to the name of any object being documented. Every object that is documented in a file needs to have a \alias entry and these names are used when users request the help page, either using the ? operator or via a call to the help function. The exact format of these depends on the type of object being documented and S4 classes and methods have a special markup.

Rd format supports marking up text in different fonts. For example, you can use \kbd to denote keyboard input and help users distinguish between what they should type and what the response will be. There are facilities for tables, sections, and for including mathematical formulae and equations. You can also cross-reference other documentation via the \link markup. For example, \code{\link{foo}}) will produce a hyperlink to the help page for object foo. You can also be more specific and use optional arguments such as \link[pkg]{foo} and \link[pkg:bar]{foo} to link to the package **pkg** with topic foo and bar, respectively. The syntax for linking to class, or method-specific documentation, across packages is:
\code{\link[arules:transactions-class]{transactions}}, which links to the class documentation.

7.3.3.1 Documenting S4 classes and methods

There are specialized prompt functions, promptMethods and promptClass, that can be used to provide template documentation files for S4 classes and methods. One strategy that seems to work reasonably well is to document all classes and generic functions with their own help pages. Documenting methods, especially when the package provides a method for a generic function that is defined in some other R package, can be problematic. The Rd system does not facilitate duplicate documentation, nor is it dynamic in the sense that documentation files cannot be updated as packages are attached or detached. These restrictions make it quite difficult to provide documentation for methods that users can find. The S4Help from the **RBioinf** package is designed to alleviate some of the problems of finding S4 documentation.

7.3.4 Name spaces

R packages are the *de facto* level at which names, or symbols, are managed. It is not possible to have two different objects that share the same name in the same package. A name space is a tool that is used to control which symbols are visible and which values are used during evaluation. Name spaces in R are described in some detail in Tierney (2003).

A name space allows the author of a package to decide what functionality to export and allow users access to and which parts of the implementation to keep private. For example, name spaces allow developers to adopt a development strategy that makes use of many helper functions, without having to document or expose those functions. Name spaces provide a clearer interface for users and state explicitly what functions are part of the API. At the same time they allow the developer to experiment with different implementations and data structures, and to make other changes that are then transparent to the end user.

Name spaces ensure that the values associated with symbols in the **base** package are not shadowed by other definitions, or bindings, unless the package author makes such changes explicitly. Further, name spaces allow access to variables and their values without necessarily attaching the corresponding package to the search path, thus reducing the overhead in searching for symbols and decreasing the probability that a variable binding is inadvertently shadowed.

With name spaces it is necessary to distinguish between loading a package and attaching a package. Loading a package makes the code and variables in it available, but does not place the package on the search path. Attaching a package both loads and attaches the package onto the search path. Initialization is discussed in Section 7.4.

A name space is added to a package by the inclusion of a `NAMESPACE` file in the top folder of the package. The `NAMESPACE` file contains a number of different directives, most of which are listed below.

export specifies which symbols are to be exported.

exportPattern a regular expression that indicates which symbols are to be exported.

exportClasses describes the S4 classes exported.

exportMethods describes the S4 methods that are exported.

import specifies package names from which all symbols are imported.

importFrom first the name of the package to import from, followed by the names of all symbols to be imported.

importMethodsFrom indicates which generic functions are imported from the named package.

importClassesFrom indicates which classes are imported from the named package.

S3method allows the author to export an S3 method. For example, S3method("write", "tlp") exports the method write.tlp and allows for dispatch to work on this S3 method.

useDynLib specifies the name of a shared object library that should be loaded and used by the package for calling C, FORTRAN or code in some other compiled language.

A name space is sealed. Sealing prevents changes to the bindings and once a package with a name space is loaded, it is not possible to add or remove variables or to change the values of variables defined in a name space. If it is important to have variables with mutable bindings, the recommended strategy is to place those variables inside an environment within the name space. The bindings in the environment can be altered, and in most cases access to them can be controlled, thereby providing a very useful mechanism for managing state information.

It is possible to obtain non-exported values from a name space but this practice should be avoided and the functionality is really intended to support debugging during development. The main functions in this area are the triple-colon operator, `:::`, so that foo:::bar says to get the value associated with the symbol bar in the name space foo. The function getFromNamespace performs a similar role and is more flexible. Other functions such as assignInNamespace and fixInNamespace provide a way to change values in a name space, the latter playing a role similar to the function fix, which allows the user to edit a function definition and replace the existing value with the edited definition.

Name spaces are implemented using R environments. A name space consists of three static frames. The first static frame contains the local definitions for the package, the second static frame contains the imports and the third static frame contains the definitions in the **base** package. The third static frame ensures that the global variables from the **base** package will not be shadowed by definitions on the search path. Developers who want to shadow definitions in **base** can do so by defining new values in their code, or by explicitly importing the symbols from some other package.

Data sets are explicitly excluded from the name space and cannot be accessed via the double-colon operator. This design decision was taken to decrease the likelihood of collisions between data sets and functions with the same names.

7.3.5 Finding out about name spaces

There are a number of functions that can be used to find out about a name space. These provide a form of reflection that allows users and developers to assess which name spaces are loaded, what they import and export, and which packages make use of a package by importing it. In the code below we load the **Biobase** package and see what it imports, then find out which name spaces are using, or relying on, the **tools** package.

```
> library("Biobase")
> getNamespaceImports("Biobase")

$base
[1] TRUE

$tools
[1] TRUE

> getNamespaceUsers("tools")

[1] "Biobase"
```

7.4 Initialization

Many package developers want to have a message printed when the package is attached. And that is perfectly reasonable for interactive use, but there can be situations where it is particularly problematic. In order to make it easy for others to suppress your start-up message, you should construct it using `packageStartupMessage` and then `suppressPackageStartupMessages` can be used to suppress the message if needed.

When the code in a package is assembled into a form suitable for use with R, via the package building system, there are some computations that can happen once, at build time, others that must happen at the time the package is installed, and still others that must happen every time the package is attached or loaded. Construction of internal class hierarchies, for S4 classes and methods, where all elements are either in the recommended packages or in the package being built, can be performed at build time. Finding and linking to system libraries must be done at install time, and in many cases again at load time.

If the function `.First.lib` is defined in a package, it is called with arguments `libname` and `pkgname` after the package is loaded and attached. While it is rare to detach packages, there is a corresponding function `.Last.lib`, which if defined will be called when a package is detached.

When a `NAMESPACE` file is present, the package can be either loaded or attached. Since there is a difference between loading and attaching the single initialization function, `.First.lib` is no longer sufficient. A package with a name space can provide two functions: `.onLoad` and `.onAttach`. These functions, if defined, should not be exported. Many packages will not need either function, since import directives take the place of calls to `require` and use-DynLib directives can replace direct calls to `library.dynam`.

When a package with a name space is supplied as an argument to the `library` function, first `loadNamespace` is invoked and then `attachNamespace`. If a package with a name space is loaded due to either an import directive or the double-colon operator, then only `loadNamespace` is called. `loadNamespace` checks whether the name space is already loaded and registered with the internal registry of loaded name spaces. If so, the loaded name space is returned, and it is not loaded a second time. Otherwise, `loadNamespace` is called on all imported name spaces, and definitions of exported variables of these packages are copied to the imports frame for the package being loaded. Then either the package code is loaded and run or the binary image is loaded, if it exists. Finally, the `.onLoad` function is run, if the package defined one.

7.4.1 Event hooks

A set of functions is available that can be used to set actions that should be performed before packages are attached or detached, and similarly before name spaces are loaded or unloaded. These functions are `getHook`, `setHook` and `packageEvent`. Among other things, these hooks allow users to have some level of control over what happens when a package is attached or detached.

Chapter 8

Data Technologies

8.1 Introduction

Handling data efficiently and effectively is an essential task in Bioinformatics. In this chapter we present some of the many tools that are available for dealing with data. The R Data Import/Export Manual (R Development Core Team, 2007a) should also be consulted for other topics and in many cases for more details regarding different technologies and their interfaces in R. The solution to many bioinformatic tasks will require some use of web-oriented technologies. Generating requests, posting and reading forms data, as well as locating and using web services are programming tasks that are likely to arise.

We begin our discussion by describing a range of tools that have been implemented in R and that can be used to process and transform data. Next we discuss the different interfaces to databases that are available but focus our discussion on SQLite as it is used extensively within the Bioconductor Project. We then discuss capabilities for interacting with data sources in XML. We conclude this chapter by considering the usage of different bioinformatic data sources via web protocols and in particular discuss some resources available from the NCBI and also demonstrate some basic usage of the **biomaRt** package.

8.1.1 A brief description of GO

GO (The Gene Ontology Consortium, 2000) is a valuable bioinformatic resource that consists of an ontology, or restricted vocabulary, of terms that are ordered as a directed acyclic graph (DAG). We will use GO as the basis for a number of examples in this chapter and hence give a brief treatment. GO is described in more detail in Gentleman et al. (2004) and Hahne et al. (2008). There are three separate components: molecular function (MF), biological process (BP) and cellular component (CC). GO uses its own set of identifiers and for each term, detailed information is available. A separate project (Camon et al., 2004) maps genes, really proteins, to GO terms. There are a number of *evidence codes* that are used to explain the reason that a gene was mapped to a particular term.

8.2 Using R for data manipulation

We have seen many of the different capabilities of R in Chapter 2. Here, we take a slightly different approach and concern ourselves mainly with data processing, that is, with those tasks that take as input one or more data sets and process them to provide a new, processed output data set that can be used for other purposes. There are many different solutions to most of these tasks, and our goal is to provide some broad coverage of the different capabilities.

We will make use of data from one of the metadata packages to demonstrate some of the different computations. Using these data, one is typically interested in counting things, such as "How many probes on a microarray correspond to genes that lie on chromosome 7?", or in dividing the probes according to chromosomal location, or selecting one probe to represent each distinct Entrez Gene ID.

8.2.1 Aggregation and creating tables

Aggregating data and computing simple summaries are common tasks, and there are specialized efficient functions for performing many of them. We first load the **hgu95av2** metadata package and then will extract the information about which chromosome each probe is located on. This is a bit cumbersome and would not be how you should approach this problem in practice, since there are other tools (see Section 8.2.2) that are more appropriate.

```
> library("hgu95av2")
> chrVec = unlist(as.list(hgu95av2CHR))
> table(chrVec)

chrVec
   1    10   11   12   13   14   15   16   17   18   19
1234  453  692  698  225  408  349  500  724  171  745
   2    20   21   22    3    4    5    6    7    8    9
 807  309  147  350  661  448  537  716  573  419  426
   X    Y
 499   41

> class(chrVec)

[1] "character"
```

Exercise 8.1
Which chromosome has the the most probe sets and which has the fewest?

Next, we might want to know the identities of those genes on the Y chromosome. We can solve this problem in many different ways, but since we might want to ultimately plot values in chromosome coordinates, we will make use of the function `split`. In the code below, we split the names of `chrVec` because they correspond to the different chromosomes. The return value is a list of length 25 where each element has all the Affymetrix probe IDs for the probes that correspond to genes located in the chromosome. We then use the `sapply` to check our results, and can compare the answer with that found above using `table`.

```
> byChr = split(names(chrVec), chrVec)
> sapply(byChr, length)
```

	1	10	11	12	13	14	15	16	17	18	19
	1234	453	692	698	225	408	349	500	724	171	745
	2	20	21	22	3	4	5	6	7	8	9
	807	309	147	350	661	448	537	716	573	419	426
	X	Y
	499	41

Then we can list all of the probe sets that are found on any given chromosome simply by subsetting `byChr` appropriately.

```
> byChr[["Y"]]
 [1] "629_at2"      "39168_at2"    "34215_at2"
 [4] "31415_at"     "40342_at"     "32930_f_at"
 [7] "31911_at"     "31601_s_at"   "35930_at"
[10] "1185_at2"     "32991_f_at"   "40436_g_at2"
[13] "36553_at2"    "33593_at"     "31412_at"
[16] "35885_at"     "35929_s_at"   "32677_at"
[19] "41108_at2"    "41138_at2"    "31414_at"
[22] "36321_at"     "38182_at"     "40097_at"
[25] "31534_at"     "40030_at"     "41214_at"
[28] "38355_at"     "32864_at"     "40435_at2"
[31] "36554_at2"    "35073_at2"    "34477_at"
[34] "31411_at"     "34753_at2"    "35447_s_at2"
[37] "32428_at"     "37583_at"     "33665_s_at2"
[40] "34172_s_at2"  "31413_at"
```

apply	matrices, arrays, data.frames
lapply	lists, vectors
sapply	lists, vectors
tapply	atomic objects, typically vectors
by	similar to `tapply`
eapply	environments
mapply	multiple values
rapply	recursive version of `lapply`
esApply	*ExpressionSets*, defined in **Biobase**

Table 8.1: Different forms of the apply functions.

8.2.2 Apply functions

There are a number of functions, listed in Table 8.1, that can be used to apply a function, iteratively, to a set of inputs. The `apply` function operates on arrays, matrices or data.frames where one would like to apply a function to each row, or each column; and in the case of arrays, to any other dimension. The notion is easily extended to lists, where `lapply` and `sapply` are appropriate, or to ragged arrays, `tapply`, or to environments, `eapply`. If the problem requires that a function be applied to two or more inputs, then `mapply` may be appropriate. When possible, the return value of the apply functions is simplified. These functions are not particularly efficient and for large matrices more efficient alternatives are discussed in Section 8.2.3. One of the main reasons to prefer the use of an apply-like function, over explicitly, using a `for` loop is that it more clearly and succinctly indicates what is happening and hence makes the code somewhat easier to read and understand.

The return value from `apply` will be simplified if possible, in that if all values are vectors of the same length, then a matrix will be returned. The matrix will have one column for each computation and one row for each value returned. Thus, if a matrix has two columns and five rows and a function that returns three values is applied to the rows, the return value will be a matrix with three rows and five columns.

The function `tapply` takes as input an atomic vector and a list of one or more vectors (usually factors) of the same length as the first argument. The first argument is split according to the unique values of the supplied factors, and the specified summary statistic is computed for those values in each group. For `tapply`, users can specify whether or not to try and simplify the return value using the `simplify` argument.

For lists, `lapply` will apply a function to each element, and it does not attempt to simplify the return value, which will always be a list of the same length as the list that was operated on. One cannot use an S4 class that is coercible to a list in a call to `lapply`; that is because S4 methods set on `as.list` will not be detected when the code is evaluated, so that the user-defined conversion will not be used. To obtain a simplified result, for example if all return values are the same length, then use `sapply`. If you want to operate on a subset of the values in the list, leaving others unchanged, then `rapply` can be used. With `rapply` you can specify the class of objects to be operated on; the other elements can either be left unchanged, or replaced by a user-supplied default value.

To apply a function to every element of an environment, use `eapply`. The order of the output is not guaranteed, as there is no ordering of values in the environment. By default, objects in the environment whose names begin with a dot are not included.

In the **Biobase** package, a function named `esApply` is provided to simplify the process of applying a function to either the rows or columns of the expression data contained within the *ExpressionSet*. It also simplifies the use of phenotypic data, from the *pData* slot in the function being applied.

8.2.2.1 An `eapply` example

The `hgu95av2MAP` contains the mappings between Affymetrix identifiers and chromosome band locations. For example, in the code below we find the chromosome band that the gene for probe `1001_at` (TIE1) maps to.

```
> library("hgu95av2")
> hgu95av2MAP$"1001_at"

[1] "1p34-p33"
```

We can extract all of the map locations for a particular chromosome or part of a chromosome by using regular expressions (Section 5.3) and the apply family of functions. Suppose we want to find all genes that map to the p arm of chromosome 17. Then we know that their map positions will all start with the characters 17p. This is a simple regular expression, ^17p, where the caret, ^, means that we should match the start of the word. We do this in two steps. First we use `eapply` and `grep` and ask for grep to return the value that matched.

```
> myPos = eapply(hgu95av2MAP, function(x) grep("^17p",
+       x, value = TRUE))
> myPos = unlist(myPos)
> length(myPos)

[1] 190
```

Exercise 8.2

Use the function ppc *from Exercise 2.16 to create a new function that can find and return the probes that map to any chromosome (just prepend the caret to the chromosome number) or the chromosome number with a p or a q after it.*

8.2.3   Efficient apply-like functions

While the apply family of functions provides a very useful abstraction and a succinct notation, its generality precludes a truly efficient implementation. For this reason there are other, more efficient functions, for tasks that are often performed, provided in R and in some add-on packages. These include rowSums, rowMeans, colSums and colMeans, which compute, per row or column, sums and means for numeric arrays. If the input array is a data.frame, then these functions first attempt to coerce it to a matrix and if successful then the operations are carried out directly on the matrix. From **Biobase**, a number of other functions for finding quantiles are implemented based on the function rowQ, which finds specified sample quantiles on a per-row basis for numeric arrays. Based on this function, other often-wanted summaries, rowMin, rowMax, etc. have been implemented.

For statistical operations, the functions rowttests, rowFtests and rowpAUCs, all from the **genefilter** package, provide very efficient tools for computing *t*-tests, F-tests and various quantities related to receiver operator curves (ROCs) in a row-wise fashion. There are methods for matrices and *ExpressionSets*.

8.2.4   Combining and reshaping rectangular data

Data in the form of a rectangular array, either a matrix or a data.frame, can be combined into new rectangular arrays in many different ways. The most commonly used functions are rbind and cbind. The first joins arrays row-wise; any number of input arrays can be specified, but they must all have the same number of columns. The row names are retained, and the column names are obtained from the first argument (left to right) that has column names. For cbind, the arrays must all have the same number of rows, and otherwise the operation is symmetric to that of rbind; an example of their use is given in the next code chunk.

```
> x = matrix(1:6, nc = 2, dimnames = list(letters[1:3],
+     LETTERS[1:2]))
> y = matrix(21:26, nc = 2, dimnames = list(letters[6:8],
+     LETTERS[3:4]))
> cbind(x, y)

  A B  C  D
a 1 4 21 24
```

```
b 2 5 22 25
c 3 6 23 26

> rbind(x, y)

    A  B
a   1  4
b   2  5
c   3  6
f  21 24
g  22 25
h  23 26
```

Data matrices with row, or column, names can be merged to form a new combined data set. The operation is similar to the *join* capability of most databases and is accomplished using function `merge`. The function supports merging data frames or matrices on the basis of either shared row or column names, as well as on other values.

In some settings it is of interest to reshape an input data matrix. This commonly arises in the analysis of repeated measures and other statistical problems, but a similar issue arises in bioinformatics when dealing with some of the different metadata resources. The function `reshape` helps to transform a data set from the representation where different measurements on the same individual are represented by different rows, to one where the different measurements are represented by columns.

One other useful function is `stack`, which can be used to concatenate vectors and simultaneously compute a factor that indicates the original input vector each value corresponds to. `stack` expects either a named list or a data frame as its first argument and, further, that each entry of that list or data frame is itself a vector. It returns a data frame with two columns and as many rows as there are values in the input, where the columns are named `values` and `ind`. The function `unstack` expects an input data frame with two columns and attempts to undo the stacking. If all vectors are the same length, then the return value from `unstack` is a data frame.

```
> s1 = list(a = 1:3, b = 11:12, c = letters[1:6])
> ss = stack(s1)
> ss

  values ind
1      1   a
2      2   a
3      3   a
```

```
4        11    b
5        12    b
6         a    c
7         b    c
8         c    c
9         d    c
10        e    c
11        f    c

> unsplit(s1, ss[, 2])

 [1] "1"   "2"   "3"   "11"  "12"  "a"   "b"   "c"   "d"   "e"
[11] "f"
```

8.3   Example

We now provide a somewhat detailed and extended example, using many of the tools and functions described to map data to chromosome bands. This information is available in all Bioconductor metadata packages; the suffix for the appropriate environment is MAP. In our example we will make use of the HG-U95Av2 GeneChip so the appropriate data environment is hgu95av2MAP.

In the next code chunk we extract the MAP locations and then carry out a few simple quality assessment procedures; we look to see if many probe sets are mapped to multiple locations and also to see how many probe sets have no MAP location.

```
> mapP = as.list(hgu95av2MAP)
> mLens = unlist(eapply(hgu95av2MAP, length))
```

Then we can use table to summarize the results.

```
> mlt = table(mLens)
> mlt

mLens
    1     2     3
12438   185     2
```

And we see that there are some oddities, in that some probe sets are annotated at several positions. There are several reasons for this. One is that there
is a homologous region shared between chromosomes X and Y, and another
is that not all gene locations are known precisely. The two probe sets that
report three locations correspond to a single gene, ACTN1. In the code below
we see that the three reported locations are all relatively near each other and
most likely reflect difficulties in mapping.

```
> len3 = mLens[mLens == 3]
> hgu95av2SYMBOL[[names(len3)[1]]]

[1] "ACTN1"

> hgu95av2MAP[[names(len3)[1]]]

[1] "14q24.1-q24.2" "14q24"         "14q22-q24"
```

Exercise 8.3
*How many genes are in the homologous region shared by chromosomes X and
Y.*

In the next example we show that there are 532 probe sets that do not have
a map position.

```
> missingMap = unlist(eapply(hgu95av2MAP,
+      function(x) any(is.na(x))))
> table(missingMap)

missingMap
FALSE  TRUE
12093   532
```

Next, we can see what the distribution of probe sets per map position is.
For those probe sets that have multiple map positions, we will simply select
the first one listed.

```
> mapPs = sapply(mapP, function(x) x[1])
> mapPs = mapPs[!is.na(mapPs)]
> mapByPos = split(names(mapPs), mapPs)
> table(sapply(mapByPos, length))

 1  2  3  4  5  6  7  8 11 13
36 26  9  5  1  1  1  1  1  1
```

Exercise 8.4
Which chromosome band has the most probe sets contained in it? How many chromosome bands are from chromosome 2? How many are on the p-arm and how many on the q-arm?

8.4   Database technologies

Relational databases are commonplace and provide a very useful mechanism for storing data in structured tables. A standard mechanism for querying relational databases is the Structured Query Language (SQL). This section is not a tutorial on either of these topics; interested readers should consult some of the very many books and other resources that describe both relational databases and SQL. Rather, we concentrate on describing the R interfaces to relational databases. R software for relational databases is covered in Section 4 of the R Data Import/Export manual R Development Core Team (2007a). There is a database special interest group (SIG) that has its own mailing list that can be used to discuss topics of interest to those using databases.

Databases for which there are existing R packages include SQLite (**RSQLite**), Postgres (**RdbiPgSQL**), MySQL (**RMySQL**) and Oracle (**ROracle**). In addition there is an interface to the Open Database Connectivity (ODBC) standard via the **RODBC** package. Most of these rely on there being an instance of the particular database already installed, **RSQLite** being an exception. Three of these packages, **RSQLite**, **RMySQL** and **ROracle**, use the DBI interface and hence depend on the **DBI** package. The Postgres driver has not been updated to the DBI interface. The **DBI** package provides a common interface to all supported databases, thereby allowing users to use their favorite database and have the R code be relatively unchanged. Ideally one should be able to write code in R that will perform well, regardless of which database engine was actually used.

There are two basic reasons to want to interact with a database from within R. One is to allow for access to large, existing data collections that are stored in a database. For these sorts of interactions, the database typically already exists and a user will need to have an account and access permission to connect to the database. The user then typically make requests for specific subsets of the data. A second common usage is for more interactive usage, where data are transferred from R to the database and vice versa. Recently the Bioconductor Project has begun moving annotation data packages into a relational database format, relying primarily on SQLite. In addition, as the size of microarray and other high throughput data sets increases, it will become problematic to retain all data in memory and database interactions will likely be used to help manage very large data sets. It will seldom be sensible to create large database tables from R. Most databases have specialized import routines that will substantially reduce the time required to install and create large tables.

8.4.1 DBI

DBI provides a number of classes and generic functions (see Chapter 3 for more details on object-oriented programming) for database interactions. Different packages then support specific implementations of methods for particular databases. Some functions are required, and every package must implement them in order to be DBI compliant; other functions are optional and packages can implement the underlying functionality or not.

In the next code chunk we demonstrate the DBI equivalent of *Hello World*, using SQLite. In this example, we attach the **SQLite** package, then initialize a DBI driver and establish a connection. For SQLite, a database can be a file on the local system; and in the code below, if the database named `test` does not exist, it will be created.

```
> library("RSQLite")
> m = dbDriver("SQLite")
> con = dbConnect(m, dbname = "test")
> data(USArrests)
> dbWriteTable(con, "USArrests", USArrests, overwrite = TRUE)

[1] TRUE

> dbListTables(con)

[1] "USArrests"
```

One of the important features of the DBI interface is the notion of a result set. The function `dbSendQuery` submits and executes the SQL statement, but does not extract any records; rather, these are retained on the database side until they are requested via a call to `fetch`. The result set remains open until it is explicitly cleared via a call to `dbClearResult`. If you forget to save the result set, it can be obtained by calling `dbListResults` on the connection object. Result sets allow users to perform queries that may result in very large data sets and still control their transfer to R. As seen in the code below, setting the parameter `n` to a negative value in a call to `fetch` retrieves all remaining records.

```
> rs = dbSendQuery(con, "select * from USArrests")
> d1 = fetch(rs, n = 5)
> d1
```

```
  row_names Murder Assault UrbanPop Rape
1   Alabama   13.2     236       58 21.2
2    Alaska   10.0     263       48 44.5
3   Arizona    8.1     294       80 31.0
4  Arkansas    8.8     190       50 19.5
5 California    9.0     276       91 40.6

> dbHasCompleted(rs)

[1] FALSE

> dbListResults(con)

[[1]]
<SQLiteResult:(2687,0,7)>

> d2 = fetch(rs, n = -1)
> dbHasCompleted(rs)

[1] TRUE

> dbClearResult(rs)

[1] TRUE
```

One can circumvent dealing with result sets by using dbGetQuery instead of
dbSendQuery. dbGetQuery performs the query and returns all of the selected
data in a data frame, dealing with fetching and result sets internally.

A call to dbListTables will show all tables in the database, regardless of the
type of database. The syntax required is quite different for different databases;
hence DBI helps to simplify some aspects of database interaction. We show
the SQLite variant below.

```
> dbListTables(con)

[1] "USArrests"

> dbListFields(con, "USArrests")

[1] "row_names" "Murder"    "Assault"   "UrbanPop"
[5] "Rape"
```

Exercise 8.5
*Is there a **DBI** generic function that will retrieve an entire table in a single*

command. If so, what is its name, and what is its return value?

A SQLite-specific way of listing all tables is given in the example below.

```
> query = paste("SELECT name FROM sqlite_master WHERE",
+     "type='table' ORDER BY name;")
> rs = dbSendQuery(con, query)
> fetch(rs, n = -1)


      name
1 USArrests
```

Exercise 8.6
Select all entries from the USArrests *database where the murder rate is larger than 10.*

8.4.2   SQLite

SQLite is a very lightweight relational database, with a number of advanced features such as the ability to store Binary Large Objects (BLOBs) and to create prepared statements. SQLite stores each database as a file, the format of which is platform independent, so these files can be moved to other computers and will work on those platforms and hence it is well suited as a method for storing large data files.

In the code below we load the **SQLite** package, initialize a driver and then open a database that has been supplied with the **RBioinf** package that accompanies this volume. The database contains a number of tables that map between identifiers on the Affymetrix HG-U95Av2 GeneChip and different quantities of interest such as GO categories or PubMed IDs (that map to published papers that discuss the corresponding genes). We then list the tables in that database.

```
> library("RSQLite")
> m = dbDriver("SQLite")
> testDB = system.file("extdata/hgu95av2-sqlite.db",
+     package = "RBioinf")
> con = dbConnect(m, dbname = testDB)
> tabs = dbListTables(con)
> tabs

[1] "acc"          "go_evi"      "go_ont"
[4] "go_ont_name"  "go_probe"    "pubmed"

> dbListFields(con, tabs[2])
```

Table Name	Description	Field Names
acc	map between Affymetrix and Genbank	affy_id, acc_num
go_evi	descriptions of evidence codes	evi, description
go_ont	map from GO ID to Ontology	go_id, ont
go_ont_name	long names of GO Ontologies	ont, ont_name
go_probe	map from Affymetrix ID to GO, with evidence codes	affy_id, go_id, evi
pubmed	map from Affymetrix IDs to PubMed IDs	affy_id, pm_id

Table 8.2: Description of the tables in the test database supplied with the **RBioinf** package.

```
[1] "evi"          "description"
```

The database has six tables and they are described in Table 8.2. The different tables map between Affymetrix identifiers, GO identifiers and labels, as well one table that maps to PubMed identifiers.

Exercise 8.7
For each table in the hgu95av2.db *database, determine the type of each field.*

Exercise 8.8
How many GO evidence codes are there, and what are they?

8.4.2.1 Inner joins

The go_ont table maps GO IDs to the appropriate GO ontology, one of BP, MF or CC. We can extract data from the go_ont_name table to get the more descriptive listing of the ontology for each GO identifier. This requires an *inner join*, which is demonstrated in the code below. We first use paste to construct the query, which will then be used in the call to dbSendQuery. The inner join is established in the WHERE clause, where we require the two references to be identical. We only fetch and show the first three results.

```
> query = paste("SELECT go_ont.go_id, go_ont.ont,",
+      "go_ont_name.ont_name FROM go_ont,",
+      "go_ont_name WHERE (go_ont.ont = go_ont_name.ont)")
> rs = dbSendQuery(con, query)
> f3 = fetch(rs, n=3)
> f3
```

```
       go_id ont            ont_name
1 GO:0004497  MF Molecular Function
2 GO:0005489  MF Molecular Function
3 GO:0005506  MF Molecular Function

> dbClearResult(rs)

[1] TRUE
```

Exercise 8.9
Use an inner join to relate GenBank IDs to GO ontology codes.

8.4.2.2 Self joins

The following compound statement selects all Affymetrix probes annotated at GO ID GO:0005737 with evidence codes IDA and ISS. This uses a *self join* and demonstrates a common abbreviation syntax for table names.

```
> query = paste("SELECT g1.*, g2.evi FROM go_probe g1,",
+     "go_probe g2 WHERE  (g1.go_id = 'GO:0005737' ",
+     "AND g2.go_id = 'GO:0005737') ",
+     "AND (g1.affy_id = g2.affy_id) ",
+     "AND (g1.evi = 'IDA' AND g2.evi = 'ISS')")
> rs = dbSendQuery(con, query)
> fetch(rs)

    affy_id      go_id evi evi
1   41306_at GO:0005737 IDA ISS
2    1069_at GO:0005737 IDA ISS
3   38704_at GO:0005737 IDA ISS
4 39501_f_at GO:0005737 IDA ISS
```

8.4.3   Using AnnotationDbi

As of release 2.2 of Bioconductor, most annotation packages have been produced using SQLite and infrastructure in the **AnnotationDbi** package. This infrastructure provides increased flexibility and makes linking various data sources simpler. The implementation provides access to data objects in the usual way, but it also provides direct access to the tables and provides a number of more powerful functions.

Before presenting examples using this package, we first digress slightly to provide details on some of the concepts that underly the **AnnotationDbi** package. First, a bimap consists of two sets of objects, the left objects and

the right objects, where the names are unique within a set. There can be
any number of links between the left objects and the right objects, and these
can be thought of as edges. The edges can be *tagged* or named. Both the left
objects and the right objects can have named attributes associated with them.
In other words, a bimap is a bipartite graph and it represents the relationships
between two sets of identifiers. Bimaps can handle one-to-one relationships, as
well as one-to-many, and many-to-many relationships. Bimaps are represented
by the *Bimap* class. An example of a bimap would be to have probe IDs as
the left keys, GO IDs as the right keys, and edges, tagged by evidence codes,
as the edges between the probe IDs and the GO IDs.

We will demonstrate the use of the **AnnotationDbi** interface using the
hgu95av2.db package. Every annotation package provides a function that
can be used to access the SQLite connection directly. The name of that
function is concatenation of the basename of the package, hgu95av2 in this
case, and the string dbconn, separated by an underscore. Name mangling
ensures that multiple databases can be attached at the same time. This
function can be used to reopen the connection to the database if needed. We
first load the database and establish a connection.

```
> library("hgu95av2.db")
> mycon = hgu95av2_dbconn()
```

You can then query the tables in the database directly. In a slight abuse of
the idea, you can conceptualize tables in the database as arrays. However, they
are really bimaps, and we can use specialized tools to extract information from
the bimap. The toTable function displays all of the information in a map that
includes both the left and right values along with any other attributes that
might be attached to those values. The left and right keys can be extracted
using Lkeys and Rkeys, respectively.

```
> colnames(hgu95av2GO)

[1] "probe_id" "go_id"      "Evidence" "Ontology"

> toTable(hgu95av2GO)[1:10, ]

    probe_id       go_id Evidence Ontology
1    1000_at GO:0006468      IDA       BP
2    1000_at GO:0006468      IEA       BP
3    1000_at GO:0007049      IEA       BP
4    1001_at GO:0006468      IEA       BP
5    1001_at GO:0007165      TAS       BP
6    1001_at GO:0007498      TAS       BP
```

```
 7  1003_s_at  GO:0006928      TAS       BP
 8  1003_s_at  GO:0007165      IEA       BP
 9  1003_s_at  GO:0007186      TAS       BP
10  1003_s_at  GO:0042113      IEA       BP

> Lkeys(hgu95av2GO)[1:10]

 [1] "1000_at"   "1001_at"   "1002_f_at" "1003_s_at"
 [5] "1004_at"   "1005_at"   "1006_at"   "1007_s_at"
 [9] "1008_f_at" "1009_at"

> Rkeys(hgu95av2GO)[1:10]

 [1] "GO:0008152" "GO:0006953" "GO:0006954" "GO:0019216"
 [5] "GO:0006928" "GO:0007623" "GO:0006412" "GO:0006419"
 [9] "GO:0008033" "GO:0043039"
```

The links function returns a data frame with one row for each link, or edge, in the bimap that it is applied to. It does not report attribute information.

```
> links(hgu95av2GO)[1:10, ]

     probe_id       go_id Evidence
1    1000_at  GO:0006468      IDA
2    1000_at  GO:0006468      IEA
3    1000_at  GO:0007049      IEA
4    1001_at  GO:0006468      IEA
5    1001_at  GO:0007165      TAS
6    1001_at  GO:0007498      TAS
7  1003_s_at  GO:0006928      TAS
8  1003_s_at  GO:0007165      IEA
9  1003_s_at  GO:0007186      TAS
10 1003_s_at  GO:0042113      IEA
```

A common programming task is to invert the mapping, which typically goes from probes, or genes, to other quantities, such as their symbol. The reversed map then goes from symbol to probe or gene ID. With the old-style annotation packages, this was most easily accomplished using the reverseSplit function from **Biobase**. But with the new database annotation packages the operations are much simpler. The revmap can be used to reverse most maps. It takes as input an instance of the *Bimap* class and returns a function that can be queried using keys from that correspond to values in the original mapping. In the example below, we reverse the map from probes to symbols and then use the returned function to find all probes associated with the symbol ABL1.

```
> is(hgu95av2SYMBOL, "Bimap")

[1] TRUE

> rmMAP = revmap(hgu95av2SYMBOL)
> rmMAP$ABL1

[1] "1635_at"   "1636_g_at" "1656_s_at" "2040_s_at"
[5] "2041_i_at" "39730_at"
```

The `revmap` function can also be used on lists and there uses the `reverseSplit` function. A simple example is shown below.

```
> myl = list(a = "w", b = "x", c = "y")
> revmap(myl)

$w
[1] "a"

$x
[1] "b"

$y
[1] "c"
```

8.4.3.1 Mapping symbols

In this section we address a more advanced topic. The material is based on, and similar to, the presentation in Hahne et al. (2008), but the problem is important and common. We want to map from gene symbols to some other form of identifier, perhaps because the symbols were obtained from a paper, or other report, and we would like to see whether we can obtain similar findings using other data sources. But since most other sources do not use symbols for mapping, we must first map the available symbols back to some identifier, such as EntrezGene ID.

The code consists of four functions, three helpers and the main function `findEGs` that maps from symbols to Entrez Gene IDs. We need to know about the table structure to write the helper functions as they are basically R wrappers around SQL statements. The `hgu95av2_dbschema` function can be used to obtain all the information about the schema.

```
> queryAlias = function(x) {
+     it = paste("('", paste(x, collapse = "', '"),
+         "'", sep = "")
+     paste("select _id, alias_symbol from alias",
+         "where alias_symbol in", it, ");")
+ }
> queryGeneinfo = function(x) {
+     it = paste("('", paste(x, collapse = "', '"),
+         "'", sep = "")
+     paste("select _id, symbol from gene_info where",
+         "symbol in", it, ");")
+ }
> queryGenes = function(x) {
+     it = paste("('", paste(x, collapse = "', '"),
+         "'", sep = "")
+     paste("select * from genes where _id in",
+         it, ");")
+ }
> findEGs = function(dbcon, symbols) {
+     rs = dbSendQuery(dbcon, queryGeneinfo(symbols))
+     a1 = fetch(rs, n = -1)
+     stillLeft = setdiff(symbols, a1[, 2])
+     if (length(stillLeft) > 0) {
+         rs = dbSendQuery(dbcon, queryAlias(stillLeft))
+         a2 = fetch(rs, n = -1)
+         names(a2) = names(a1)
+         a1 = rbind(a1, a2)
+     }
+     rs = dbSendQuery(dbcon, queryGenes(a1[,
+         1]))
+     ans = merge(a1, fetch(rs, n = -1))
+     dbClearResult(rs)
+     ans
+ }
```

The logic is to first look to see if the symbol is current, and if not, to then
search the alias table to see if there are other less current symbols. Each
of the first two queries within the findEGs function returns the symbol (the
second columns of a1 and a2) and an identifier that is internal to the SQLite
database (the first columns). The last query uses those internal IDs to extract
the corresponding Entrez Gene IDs.

```
> findEGs(mycon, c("ALL1", "AF4", "BCR", "ABL"))

  _id symbol gene_id
1  20    ABL      25
2 540    BCR     613
3 3758   AF4    4299
4 3921   ABL    4547
```

The three columns in the return are the internal ID, the symbol and the Entrez Gene ID (`gene_id`).

8.4.3.2 Combining data from different annotation packages

By using a real database to store the annotation data, we can take advantage of its capabilities to combine data from different annotation packages, or indeed from any SQLite database. Being able to select items from multiple tables does rely on their being a common value that can be used to identify those entries that are the same. It is important to realize that the internal IDs used in the **AnnotationDbi** packages cannot be used to map between packages.

In the example here, we join tables from the **hgu95av2.db** package and the **GO.db** package. And we use GO identifiers as the link across the two data packages. We attach the GO database to the HG-U95Av2 database, but could just as well have done it the other way around. In this section we are using the term *attach* to mean attaching using the SQL function ATTACH, not the R function, or concept, of attaching. We rely on some knowledge of where the GO database is located and its name, together with the `system.file` function, to construct the path to that database. The `hgu95av2.db` package is already attached and we now use the connection to it, `mycon`, to pass the SQL query that will attach the two databases.

```
> GOdbloc = system.file("extdata", "GO.sqlite", package="GO.db")
> attachSql = paste("ATTACH '", GOdbloc, "' as go;", sep = "")
> dbGetQuery(mycon, attachSql)

NULL
```

Next, we are going to select some data, based on the GO ID, from two tables, one in the HG-U95Av2 database and one in the GO database. We limit the query to ten values. The WHERE clause on the last line of the SQL query is the part of the query that requires the GO identifiers be the same. The other parts of the query, the first five lines, set up what variables to extract and what to name them.

```
> sql = paste("SELECT DISTINCT a.go_id AS 'hgu95av2.go_id',",
+              "a._id AS 'hgu95av2._id',",
+              "g.go_id AS 'GO.go_id', g._id AS 'GO._id',",
+              "g.ontology",
+              "FROM go_bp_all AS a, go.go_term AS g",
+              "WHERE a.go_id = g.go_id LIMIT 10;")
> dataOut = dbGetQuery(mycon, sql)
> dataOut
```

	hgu95av2.go_id	hgu95av2._id	GO.go_id	GO._id
1	GO:0000002	255	GO:0000002	13
2	GO:0000002	1633	GO:0000002	13
3	GO:0000002	3804	GO:0000002	13
4	GO:0000002	4680	GO:0000002	13
5	GO:0000003	41	GO:0000003	14
6	GO:0000003	43	GO:0000003	14
7	GO:0000003	81	GO:0000003	14
8	GO:0000003	83	GO:0000003	14
9	GO:0000003	104	GO:0000003	14
10	GO:0000003	105	GO:0000003	14

	ontology
1	BP
2	BP
3	BP
4	BP
5	BP
6	BP
7	BP
8	BP
9	BP
10	BP

The query combines the go_bp_all table from the HG-U95Av2 database with the go_term table from the GO database, based on the go_id. For illustration purposes, internal ID (_id) and the go_id from both tables are included in the output. This makes it clear that the go_ids can be used to join these tables but the internal IDs cannot. The internal IDs, _id, are suitable for joins within a single database but cannot be used across databases.

8.4.3.3 Metadata about metadata

In order to appropriately combine tables from various databases, users are encouraged to look at the standard schema definitions. The latest schemas are the 1.0 schemas, and these can be found in the inst/DBschemas/schemas_1.0

directory of the **AnnotationDbi** package. These schemas can also be obtained interactively using the corresponding dbschema function, as shown below. Because all output is merely cated to the screen, we use capture.output to collect it and print only the first few tables.

```
> schema = capture.output(hgu95av2_dbschema())
> head(schema, 18)

  [1] "--"

  [2] "-- HUMANCHIP_DB schema"

  [3] "-- ===================="

  [4] "--"

  [5] ""

  [6] "-- The \"genes\" table is the central table."

  [7] "CREATE TABLE genes ("

  [8] "  _id INTEGER PRIMARY KEY,"

  [9] "  gene_id VARCHAR(10) NOT NULL UNIQUE          -- Entre
z Gene ID"
 [10] ");"

 [11] ""

 [12] "-- Data linked to the \"genes\" table."

 [13] "CREATE TABLE probes ("

 [14] "  probe_id VARCHAR(80) PRIMARY KEY,            -- manuf
acturer ID"
 [15] "  accession VARCHAR(20) NULL,                 -- GenBa
nk accession number"
 [16] "  _id INTEGER NULL,                           -- REFER
ENCES genes"
 [17] "  FOREIGN KEY (_id) REFERENCES genes (_id)"

 [18] ");"
```

The above example prints the schema used for the HG-U95Av2 database into your R session. Each database has three tables that describe the contents of that database, as well as where the information contained in the database originated. The `metadata` table describes the package itself and gives information such as the schema name, schema version, chip name and a manufacturer URL. This schema information is useful for telling users which version of the schema they should consult if they want to make queries that join different databases together, like the compound query described above. The `map_metadata` table lists the various maps provided by the package and where the data for each map was obtained. And finally, the `map_counts` table gives the number of values that are contained in that map.

A summary of the tables, number of elements that are mapped, information on the schema, and on the data used to create the package are printed by calling a function that has the same name as the package.

```
> qcdata = capture.output(hgu95av2())
> head(qcdata, 20)

 [1] "Quality control information for hgu95av2:"
 [2] ""
 [3] ""
 [4] "This package has the following mappings:"
 [5] ""
 [6] "hgu95av2ACCNUM has 12625 mapped keys (of 12625 keys)"
 [7] "hgu95av2ALIAS2PROBE has 36833 mapped keys (of 36833 keys)"
 [8] "hgu95av2CHR has 12117 mapped keys (of 12625 keys)"
 [9] "hgu95av2CHRLENGTHS has 25 mapped keys (of 25 keys)"
[10] "hgu95av2CHRLOC has 11817 mapped keys (of 12625 keys)"
[11] "hgu95av2ENSEMBL has 11156 mapped keys (of 12625 keys)"
[12] "hgu95av2ENSEMBL2PROBE has 8286 mapped keys (of 8286 keys)"
[13] "hgu95av2ENTREZID has 12124 mapped keys (of 12625 keys)"
[14] "hgu95av2ENZYME has 1957 mapped keys (of 12625 keys)"
[15] "hgu95av2ENZYME2PROBE has 709 mapped keys (of 709 keys)"
[16] "hgu95av2GENENAME has 12124 mapped keys (of 12625 keys)"
[17] "hgu95av2GO has 11602 mapped keys (of 12625 keys)"
[18] "hgu95av2GO2ALLPROBES has 8383 mapped keys (of 8383 keys)"
[19] "hgu95av2GO2PROBE has 5898 mapped keys (of 5898 keys)"
[20] "hgu95av2MAP has 12093 mapped keys (of 12625 keys)"
```

Alternatively, the contents of the `map_counts` table can be obtained from the `MAPCOUNTS` object, while the contents of the `metadata` table can be obtained by calling the appropriate `dbInfo` function, as demonstrated below.

```
> hgu95av2MAPCOUNTS
> hgu95av2_dbInfo()
```

8.4.3.4   Making new data packages with SQLForge

Included in the **AnnotationDbi** package is a collection of functions that
can be used to make new microarray annotation packages. Making a chip
annotation package is a two-step process. In simple terms, a file containing the
mapping between the chip identifiers and some standard biological identifiers
is used, in conjunction with a special intermediate database, to construct a
chip-specific database. The second step wraps that chip-specific database into
an R package.

In more detail, the first step is to construct an SQLite database that con-
forms to a schema for the organism that the chip is designed for. Conforming
to a standard schema is essential as it allows the new package to integrate
with all other annotation packages, such as **GO.db** and **KEGG.db**. This
database building step requires two inputs. It requires an input file that maps
probe IDs to another known ID, typically a tab delimited file. If the chip is
an Affymetrix chip and you have one of their csv files, you can use that as
an input instead. If a tab delimited file is used, then this file must have two
columns, where the first column is the probe ID and the second column is the
other ID and no header should be used; the first line in the file should be the
first pair of mappings. The other ID can be an Entrez Gene ID, a RefSeq ID,
a Gene Bank ID, a Unigene ID or a mixture of Gene Bank and RefSeq IDs. If
there is other information in the form of alternate IDs that are also matched
to the probe IDs, these can also be included as other, optional, files.

The second required input is an intermediate database. This database contains information for all genes in the model organism, and many different
biological entities, such as Entrez Gene, KEGG, GO, and Uniprot. These
databases are provided as Bioconductor packages and there is one package
for each supported model organism. These packages are very large, and are
not required unless you want to make annotation packages for the organism
in question. Packages can be downloaded using biocLite, as is shown below for the intermediate package needed to construct annotation for human
microarrays.

```
> source("http://bioconductor.org/biocLite.R")
> biocLite("human.db0")
```

For demonstration purposes, a file mapping probes on the HG-U95Av2
GeneChip to GenBank IDs is provided in the extdata directory of

AnnotationDbi. In the example below, we first obtain the path to that file and then set up the appropriate metadata. Details on what terms to use for each of the model organisms are given in the vignette for the **AnnotationDbi** package.

```
> hgu95av2_IDs = system.file("extdata",
                             "hgu95av2_ID",
                             package="AnnotationDbi")
> #Then specify some of the metadata for my database
> myMeta = c("DBSCHEMA" = "HUMANCHIP_DB",
      "ORGANISM" = "Homo sapiens",
      "SPECIES" = "Human",
      "MANUFACTURER" = "Affymetrix",
      "CHIPNAME" = "Affymetrix Human Genome U95 Set Version 2",
      "MANUFACTURERURL" = "http:www.affymetrix.com")
```

We then create a temporary directory to hold the database, and construct one.

```
> tmpout = tempdir()
> popHUMANCHIPDB(affy = FALSE, prefix = "hgu95av2Test",
      fileName = hgu95av2_IDs, metaDataSrc = myMeta,
      baseMapType = "gb", outputDir = tmpout,
      printSchema = TRUE)
```

In the above example, setting the DBSCHEMA value is especially important as it specifies the schema to be used for the database. The function popHUMANCHIPDB actually populates the database and its name reflects the schema that it supports. To create a mouse chip package, you would use popMOUSECHIPDB.

The second phase of making an annotation data package is wrapping these SQLite databases into an **AnnotationDbi**-compliant source package. We need to specify the schema, PkgTemplate, the version number, Version, as well as other details. Once that has been done, the function makeAnnDbPkg is used to carry out the computations and its output is a fully formed R package that can be installed and used by anyone.

```
> seed <- new("AnnDbPkgSeed", Package = "hgu95av2Test.db",
      Version = "1.0.0", PkgTemplate = "HUMANCHIP.DB",
      AnnObjPrefix = "hgu95av2Test")
> makeAnnDbPkg(seed, file.path(tmpout, "hgu95av2Test.sqlite"),
      dest_dir = tmpout)
```

```
Creating package in /tmp/RtmpvORzTT/hgu95av2Test.db
```

In order to simplify the process, there is a wrapper function that performs both steps; it makes the intermediate SQLite database and then constructs a complete annotation package. In most cases this would be preferred to the two-step option previously discussed.

```
> makeHUMANCHIP_DB(affy=FALSE,
      prefix="hgu95av2",
      fileName=hgu95av2_IDs,
      baseMapType="gb",
      outputDir = tmpout,
      version="2.1.0",
      manufacturer = "Affymetrix",
      chipName = "Affymetrix Human Genome U95 Set Version 2",
      manufacturerUrl = "http://www.affymetrix.com")
```

Functions are available for six of the major model organisms: makeHUMANCHIP_DB, makeMOUSECHIP_DB, makeRATCHIP_DB, makeFLYCHIP_DB, makeYEASTCHIP_DB, makeARABIDOPSISCHIP_DB.

8.5   XML

The eXtensible Markup Language (XML) is a widely used standard for marking up data and text in a structured way. It is an important tool for communication between different clients and servers on the World Wide Web, where servers and clients use XML dialects to negotiate queries on service availability, to submit requests, and to encode the results of requested computations. Readers unfamiliar with XML should consult one of the many references available, such as Skonnard and Gudgin (2001) or Harold and Means (2004). There are many other related tools and paradigms that can be used to interact with XML documents. In particular, XSL, which is a stylesheet language for XML, and XSL Transformations (XSLT, http://www.w3.org/TR/xslt), which can be used to transform XML documents from one form to another. The **Sxslt** package, available from Omegahat, provides an interface to an XSLT translator. Also of interest is the XPath language (http://www.w3.org/TR/xpath), which was designed for addressing parts of an XML document.

An XML document is tree-like in structure. It consists of a series of *elements*, which we will sometimes also refer to as nodes. An example is given in Program 8.1. Normally each element has both an opening tag and a closing tag, but in some circumstances these can be collapsed into a single tag. The syntax for an opening tag is to have the *name* of the element enclosed between a less-than sign, <, and a greater-than sign, >. Following the element name, and before the closing >, there can be any number of named attributes. The end of the element is signaled using a similar syntax except that here the name of the node is preceded by a forward slash. Between the opening and closing tags there can be other XML elements, plaintext, or a few other constructs, but we will only consider the simple case of plaintext here and refer the reader to the specialized references already mentioned for more details on the XML format. The most basic value of an XML element is the sub-document rooted at that element. When an element has no XML children, the value is the textual content between the opening and closing tag.

A small extract from one of the files supplied by the IntAct database (Kerrien et al., 2006) is shown in Program 8.1. In this example the first element is named `participantList` and that element has no attributes, but one child, `participant`, which has an attribute named `id`. The `participant` element itself has a subelement, named `interactorRef`. The `interactorRef` element has no attributes but it does have a value, in this case the number 803.

```
<participantList>
 <participant id="807">
  <interactorRef>803</interactorRef>
 </participant>
 ...
</participantList>
```

Program 8.1: An XML snippet.

Typically, but not necessarily, a *schema* or DTD is used to describe a specific XML format. The schema describes the allowable tags and provides information on what content is allowed. XML documents are strictly hierarchical and essentially tree-like. Each element in the document may have one or more child elements. The schema describes the set of required and allowed child elements for any given element. The only real benefit to using XML over any other form of markup is that there are good parsers, validators and many other tools that have been written in just about every computer language that can be used; neither you nor those working with you will need to write that sort of low level code.

XML name spaces are described in most books on XML as well as at `http://www.w3.org/TR/REC-xml-names/`. The use of name spaces allows the reuse of tags in different contexts. A simple example of a name space, taken from the web site named above is shown below. In this example there are two name spaces, `bk` and `isbn`. These can then be used as a prefix on the tag names in the document, e.g., `bk:title` and `isbn:number`.

```
<?xml version="1.0"?>
<!-- both namespace prefixes are available throughout -->
<bk:book xmlns:bk='urn:loc.gov:books'
         xmlns:isbn='urn:ISBN:0-395-36341-6'>
    <bk:title>Cheaper by the Dozen</bk:title>
    <isbn:number>1568491379</isbn:number>
</bk:book>
```

Program 8.2: An example of a name space in XML.

There are two basic methods for parsing XML documents. One is the document object model, or DOM, parsing and the other is event-style, or SAX, parsing. The DOM approach is to read the entire document into memory and to operate on it as a whole. The SAX approach is to parse the document and to act on different entities as the document is parsed. SAX parsing can be much more efficient for large files as it is seldom necessary to have the entire document in memory at one time.

8.5.1 Simple XPath

Our tutorial on XPath is very brief, and is intended only to make the further examples comprehensible. Readers should consult a more definitive reference if they are going to make extensive use of XPath.

//participant selects all `participant` elements.

//participant/interactorRef selects all `interactorRef` elements that are children of a `participant` element.

//@id selects all attributes that are named `id`.

//participant[@id] selects all `participant` elements that have an attribute named `id`.

There are many more capabilities; elements can be selected on the value of an attribute, the first child of an element, the last child, children with specific properties and many other criteria can be used.

8.5.2 The XML package

Processing of XML documents in R can be done with the **XML** package. The package is extensive and provides interfaces to most tools that you will need to efficiently process XML documents. While the **XML** package has relatively little explicit documentation, it has a very large number of examples that can be studied to get a better idea of the extensive capabilities that are contained in that package. These files can be found in the `examples` subdirectory for the installed package.

Given the structure of an XML document, a convenient and flexible way to process the document is to process each element with a function that is specialized to deal with that particular element. Such functions are sometimes referred to as *handlers*, and this is the approach that the **XML** package has taken.

DOM-style parsing is done by the `xmlTreeParse` function while SAX, or stream, style parsing is handled by the `xmlEventParse` function. In either case, handler functions can be defined and supplied. These handlers are then invoked, when appropriate, and facilitate appropriate processing of the document.

With `xmlTreeParse`, the value returned by the handler is then placed into the saved version of the XML document. A rather convenient way to remove elements is to provide a handler that returns `NULL`. The return value from `xmlTreeParse` is an object of class *XMLDocument*. The return value for `xmlEventParse` is the handler's argument that was supplied. It is presumed that this is a closure, and that the handlers have processed the elements and stored the quantities of interest in local variables that can be extracted later.

8.5.3 Handlers

Basically, handlers can be specified for any type of element and when a element is encountered, during parsing, the appropriate handler is invoked. Handlers can be specialized to either the start of element processing, or they can be run after the element has been processed, but before the next element is read.

For DOM processing using `xmlTreeParse`, either a named list of handlers or a single function can be supplied. If a single function is supplied, then it is called on every element of the underlying DOM tree.

For SAX processing, `xmlEventParse`, the handlers are more extensive. The standard function or handler names are `startElement`, `endElement`, `comment`, `getEntity`, `entityDeclaration`, `processingInstruction`, `text`, `cdata`, `startDocument` and `endDocument`. In addition you can provide handler functions for specific tags such as `<myTag>` by giving the handler the name `myTag`.

8.5.4   Example data

An XML file containing data from the IntAct database (Kerrien et al., 2006) is supplied as part of the **RBioinf** package. We will use it for our XML parsing examples. More extensive examples, and a reasonably complete solution for dealing with IntAct, is provided in the **RIntact** package (Chiang et al., 2007).

In the code below we find the location of that file, so that it can be parsed.

```
> Yeastfn = system.file("extdata", "yeast_small-01.xml",
      package = "RBioinf")
```

No matter what type of processing you do, you will need to ascertain some basic facts about the XML document being processed. Among these is finding out what name spaces are being used as this will be needed in order to properly process the document. This information can be obtained using the `xmlNamespaceDefinitions` function. A default name space has no name, and can be retrieved using the `getDefaultNamespace` function. We save the default name space in a variable named `namespace` and pass that to any function that needs to know the default name space. In the code below we read in the document and then ascertain whether it is using a name space, and if so what it is. This will be important for parsing the document.

```
> yeastIntAct = xmlTreeParse(Yeastfn)
> nsY = xmlNamespaceDefinitions(xmlRoot(yeastIntAct))
> ns = getDefaultNamespace(xmlRoot(yeastIntAct))
> namespaces = c(ns = ns)
```

Exercise 8.10
How many name space definitions are there for the XML document that was parsed? What are the URIs for each of them?

8.5.5   DOM parsing

DOM-style parsing basically retrieves the entire XML document into memory. It can then be processed in different ways. The simplest way to invoke DOM-style parsing is to use the `xmlTreeParse` function, which was done above. This results in a large object that is stored as a list. Since we are not interested in all of the contents of this file, we can specify handlers for the elements that are not of interest and drop them. Since the default return value is the handlers, you must be sure to ask for the tree to be returned. In the code below we remove all elements named `sequence`, `organism`, `primaryRef`, `secondaryRef` and `names`. We see that the resulting document is much smaller.

```
> nullf = function(x, ...) NULL
> yeast2 = xmlTreeParse(Yeastfn,
      handlers = list(sequence = nullf,
      organism = nullf, primaryRef = nullf,
      secondaryRef = nullf,
      names = nullf), asTree=TRUE)
```

We can easily compare the size of the two documents and see that the second is much smaller.

```
> object.size(yeastIntAct)

[1] 47253568

> object.size(yeast2)

[1] 11793648
```

If instead, it is desirable to obtain the entire tree so that it can later be manipulated and queried, it may be beneficial to set the argument useInternalNodes to TRUE. This causes the document to be stored in an internal, to the XML package, format rather than to convert it into an R object. This tends to be faster, since no conversion is done, but also restricts the sort of processing that can be done. However, one can use XPath expressions via the function getNodeSet to process the data.

Since the XML file is using a default name space, we must use it when referencing the elements in any call to getNodeSet. We make use of the namespaces object created above. Recall that it has a value named ns, and that is used to replace the ns in "//ns:attributeList" with the appropriate URI. Recall from Section 8.5.1 that the XPath directive selects all elements in the document named attributeList.

```
> yeast3 = xmlTreeParse(Yeastfn, useInternalNodes = TRUE)
> f1 = getNodeSet(yeast3, "//ns:attributeList", namespaces)
> length(f1)

[1] 10
```

We see that there are ten elements named attributeList in the document. But our real goal is to find all protein-protein interactions and we next attempt to do just that. As in almost all cases of interacting with XML files, we need to know a reasonable amount about how the document is actually constructed to be able to extract the relevant information. We will make use of both the `xmlValue` function and the `xmlAttr` functions to extract values from the elements.

We first obtain the interaction detection methods; in this case all interactions are *two hybrid*, which is a method for detecting physical protein interactions (Fields and Song, 1989). We then obtain the name of the organism being studied, *Saccharomyces cerevisiae*.

```
> iaM = getNodeSet(yeast3,
      "//ns:interactionDetectionMethod//ns:fullName",
      namespaces)
> sapply(iaM, xmlValue)

[1] "two hybrid"

> f4 = getNodeSet(yeast3, "//ns:hostOrganism//ns:fullName",
      namespaces)
> sapply(f4, xmlValue)

[1] "Saccharomyces cerevisiae"
```

We can obtain the interactors and the interactions in a similar way. We first use `getNodeSet` and an XPath specification to obtain the elements of the XML document that contain the information we want, and then can use `sapply` and `xmlValue` to extract the quantities of interest. In this case we first obtain the interactors and then the interactions they participate in.

```
> interactors = getNodeSet(yeast3,
      "//ns:interactorList//ns:interactor",
      namespaces)
> length(interactors)

[1] 503

> interactions = getNodeSet(yeast3,
      "//ns:interactionList/ns:interaction",
      namespaces)
> length(interactions)

[1] 524
```

There are 503 different interactors that are involved in 524 different interactions.

An alternative is to use xpathApply to perform both operations in a single operation.

```
> interactors = xpathApply(yeast3,
      "//ns:interactorList//ns:interactor",
      xmlValue, namespaces = namespaces)
```

A similar, but different functionality is provided by xmlApply. The function operates on the children of the node passed as an argument.

8.5.6 XML event parsing

We now discuss using the **XML** package to perform event-based parsing of an XML document. One of the advantages of this approach is that the entire document does not need to be processed and stored in R, but rather, it is processed an element at a time. To take full advantage of the event parsing model, we will rely on lexical scope (Section 2.13).

In the code below we create a few simple functions for parsing different elements in the XML file. The name of the function is irrelevant and can be whatever you want. The functions should take two arguments; the first will be the name of the element, the second will be the XML attributes. In the code below we define three separate handlers and in this example they are essentially unrelated to each other. The first one is called entSH and it first prints out the name of the element and then saves the values of two attributes, level and minorVersion. We have it print as a debugging mechanism that allows us to be sure that nodes are being handled. Then we create an environment that will be used to store these values and make it the environment of the function entSH.

```
> entSH = function(name, attrs, ...) {
      cat("Starting", name, "\n")
      level <<- attrs["level"]
      minorVersion <<- attrs["minorVersion"]
  }
> e2 = new.env()
> e2$level = NULL
> e2$minorVersion = NULL
> environment(entSH) = e2
```

In the next code, we create two more handlers, one to extract the taxonimic ID and the other to count the number of interactions. They share an environment, but that is only for expediency; there is no data sharing in the example.

```
> hOrg = function(name, attrs, ...) {
      taxid <<- c(attrs["ncbiTaxId"], taxid)
  }
> e3 = new.env()
> e3$taxid = NULL
> environment(hOrg) = e3
> hInt = function(name, attrs, ...) numInt <<- numInt +
      1
> e3$numInt = 0
> environment(hInt) = e3
```

And now we can use these handlers to parse the data file and then print out the values. The name of the handler function is irrelevant since the link between XML element name and any particular handler function is determined by the name used in the **handlers** list. So, in the code chunk below, we have installed handlers for three specific types of elements, namely **entrySet**, **hostOrganism** and **interactor**.

```
> s1 = xmlEventParse(Yeastfn, handlers = list(entrySet = entSH,
      hostOrganism = hOrg, interactor = hInt))

Starting entrySet

> environment(s1$entrySet)$level

level
  "2"

> environment(s1$hostOrganism)$taxid

ncbiTaxId
  "4932"

> environment(s1$interactor)$numInt

[1] 503
```

8.5.7 Parsing HTML

HTML is another markup language and, if machine generated, can often be parsed automatically, however, many HTML documents do not have closing tags or use non-standard markup, which can make parsing very problematic. The function `htmlTreeParse` can be used to parse HTML documents. This function can be quite useful for some *screen-scraping* activities.

For example, we can use `htmlTreeParse` to parse the Bioconductor build reports (see code example below). The return value is a list of length three and the actual HTML has been converted to XML in the `children` sublist. This can now be processed using the standard XML tools discussed previously. In the call to `htmlTreeParse`, we set `useInternalNodes` to `TRUE` so that we will be able to use XPath syntax to extract elements of interest.

```
> url = paste("http://www.bioconductor.org/checkResults/",
      "2.1/bioc-LATEST/", sep = "")
> s1 = htmlTreeParse(url, useInternalNodes = TRUE)
> class(s1)

[1] "XMLInternalDocument"
```

For example, we can extract all of the package names. To do this, we will use the `getNodeSet` function, together with XPath syntax to quickly identify the elements we want. By looking at the source code for the check page, we see that the packages are listed as the value of an element and in particular the syntax is

```
<a href="/packages/2.1/bioc/html/spikeLI.html">spikeLI</a>
```

so we first look for elements in the tree that are named `A` and have an `href` attribute. This is done using XPath syntax in the first line of the code chunk (note that currently it seems that XML translates all element names to lower case; so you must specify them in lower case). We see that there are many more such elements than there are packages, so we will need to do some more work to retrieve the package names.

```
> f1 = getNodeSet(s1, "//a[@href]")
> length(f1)

[1] 4243
```

There are two different approaches that can be taken at this point. One is to use the function `xmlGetAttr` to retrieve the values for the `href` attributes; these can then be processed by `grep` and `sub` to find those that refer to Bioconductor packages. A second approach is to return to the HTML source and there we notice that the elements we are interested in are always subelements of b elements. In the code below we refine our XPath query to select only those a elements that are direct descendants of b elements.

```
> f2 = getNodeSet(s1, "//b/a[@href]")
> p2 = sapply(f2, xmlValue)
> length(p2)

[1] 261

> p2[1:10]

 [1] "lamb1"       "wilson2"    "wellington" "liverpool"
 [5] "lemming"     "pitt"       "A"          "ABarray"
 [9] "aCGH"        "ACME"
```

We can compare our results to the web page and see that this procedure has indeed largely retrieved the package names as desired. While the process requires some manual intervention, using `htmlTreeParse` and tools provided with the **XML** package greatly simplifies the process of retrieving values from HTML, when that is necessary.

Exercise 8.11
Carry out the first suggestion above. That is, starting with f1, retrieve the element attributes and then process them via grep and gsub to find the names of the packages. Compare your results with those above.

8.6   Bioinformatic resources on the WWW

Many bioinformatic resources provide support for the remote automated retrieval of data. Several different mechanisms are used, including SOAP, responses to queries (often in XML) and the BioMart system. Examples of service providers are the NCBI, the Kyoto Encyclopedia of Genes and Genomes (KEGG) and Ensemble. In this section we discuss and provide some simple examples that make use of these resources. In most cases there are specific R packages that provide and support the required interface.

8.6.1 PubMed

The National Library of Medicine (NLM) provides support for a number of different web services. We have developed a set of tools that can be used to query PubMed. The software is contained in the **annotate** package, and more details and documentation are provided as part of that package. Some of our earlier work in this area was reported in Gentleman and Gentry (2002).

Some functions in **annotate** that provide support for accessing online data resources are itemized below.

genbank users specify GenBank identifiers and can request the related links to be rendered in the browser or returned in XML.

pubmed users specify PubMed identifiers and can request them to be rendered in the browser or returned in XML.

pm.getabst the abstracts for the specified PubMed IDs will be downloaded for processing.

pm.abstGrep supports processing downloaded abstracts via grep to find terms contained in the abstract, such as the name of your favorite gene.

8.6.2 NCBI

In this example, we initiate a request to the EInfo utility to provide a list of all the databases that are available through the NCBI system. These can then be queried in turn to determine what their contents are. And indeed, it is possible to build a system, in R, for querying the NCBI resources that would largely parallel the functionality supplied by the **biomaRt** package, which is discussed in some detail in Section 8.6.3.

```
> ezURL = "http://eutils.ncbi.nlm.nih.gov/entrez/eutils/"
> t1 = url(ezURL, open = "r")
> if (isOpen(t1)) {
      z = xmlTreeParse(paste(ezURL, "einfo.fcgi",
          sep = ""), isURL = TRUE, handlers = NULL,
          asTree = TRUE)
      dbL = xmlChildren(z[[1]]$children$eInfoResult)$DbList
      dbNames = xmlSApply(dbL, xmlValue)
      length(dbNames)
      dbNames[1:5]
  }

      DbName        DbName        DbName        DbName
    "pubmed"    "protein" "nucleotide"      "nuccore"
      DbName
    "nucgss"
```

We see that at the time the query was issued, there were 37 databases. The names of five of them are listed, and the others can be retrieved from the `dbNames` object. Parsing of the XML is handled by fairly standard tools, and in particular we want to draw attention to the *apply*-like functions. Because XML document objects have complex R representations, the use of XPath and `xmlApply` will generally simplify the code that needs to be written.

8.6.3   biomaRt

BioMart is a query-oriented data system that is being developed by the European Bioinformatics Institute (EBI) and the Cold Spring Harbor Laboratory (CSHL). The **biomaRt** package provides an interface to BioMart.

```
> library("biomaRt")
> head(listMarts())

                        biomart
1                       ensembl
2     compara_mart_homology_49
3 compara_mart_pairwise_ga_49
4 compara_mart_multiple_ga_49
5                           snp
6               genomic_features
                                    version
1             ENSEMBL 49 GENES (SANGER)
2          ENSEMBL 49 HOMOLOGY (SANGER)
3 ENSEMBL 49 PAIRWISE ALIGNMENTS (SANGER)
4 ENSEMBL 49 MULTIPLE ALIGNMENTS (SANGER)
5         ENSEMBL 49 VARIATION  (SANGER)
6    ENSEMBL 49 GENOMIC FEATURES (SANGER)
```

Users can then select one of the BioMart databases to query; we will select the **ensembl** mart. We can then query that mart to find out which data sets it supports, and we will choose to use the human one.

```
> ensM = useMart("ensembl")
> ensData = head(listDatasets(ensM))
> dim(ensData)

[1] 6 3

> ensMH = useDataset("hsapiens_gene_ensembl", mart = ensM)
```

If you know both the name of the BioMart server and data set in advance, you can make the whole request in one step.

```
> ensMH = useMart("ensembl",
    dataset = "hsapiens_gene_ensembl")
```

Now we are ready to make data requests. **biomaRt** supports many more interactions than we will be able to cover, so interested readers should refer to the package vignette for more details and examples.

To understand **biomaRt**'s query API, we must understand what the terms *filter* and *attribute* mean. A filter defines a restriction on a query; for example, you might obtain results for a subset of genes, filtered by a gene identifier. Attributes define the values we want to retrieve, for instance, GO identifiers; or PFAM identifiers for the selected genes. You can get a listing of available filters with `listFilters` and a listing of the available attributes with `listAttributes`.

```
> filterSummary(ensMH)

  category                        group
1  FILTERS                        GENE:
2  FILTERS                  EXPRESSION:
3  FILTERS                      REGION:
4  FILTERS               GENE ONTOLOGY:
5  FILTERS                     PROTEIN:
6  FILTERS                         SNP:
7  FILTERS MULTI SPECIES COMPARISONS:

> lfilt = listFilters(ensMH, group = "GENE:")
> nrow(lfilt)

[1] 166

> head(lfilt)

                      name             description
1           affy_hc_g110       Affy hc g 110 ID(s)
2         affy_hc_g110-2       Affy hc g 110 ID(s)
3         affy_hg_focus       Affy hg focus ID(s)
4       affy_hg_focus-2       Affy hg focus ID(s)
5   affy_hg_u133_plus_2 Affy hg u133 plus 2 ID(s)
6 affy_hg_u133_plus_2-2 Affy hg u133 plus 2 ID(s)
```

We can see that there are several types of filters. There are two filters in the GENE group. We next query the attributes to see which ones are available.

```
> head(attributeSummary(ensMH))

  category            group
1 Features         EXTERNAL:
2 Features             GENE:
3 Features       EXPRESSION:
4 Features          PROTEIN:
5 Features  GENOMIC REGION:
6 Homologs AEDES ORTHOLOGS:

> lattr = listAttributes(ensMH, group = "PROTEIN:")
> lattr

                           name                 description
1                        family           Ensembl Family ID
2            family_description          Family Description
3                      interpro                Interpro ID
4          interpro_description       Interpro Description
5    interpro_short_description Interpro Short Description
6                          pfam                    PFAM ID
7                        prints                  PRINTS ID
8                       prosite                 Prosite ID
9                    prot_smart                   SMART ID
10                signal_domain              Signal domain
```

8.6.3.1 A small example

We will begin with a small set of three Entrez Gene IDs: 983 (CDC2), 3581 (IL9R) and 1017 (CDK2). The function getGene can be used to retrieve the corresponding records. Note that it returns one record per Ensembl transcript ID, which is often more than the number of Entrez Gene IDs. In the code below, we use getGene to retrieve gene-level data, and print out the symbols for the three genes. Note that the order of the genes in the return value need not be in the same order as in the request. Also, the getGene interface provides a limited set of values; if you want more detailed information, you will need to use getBM and the attributes and filters, described above, or one of the other helper functions in **biomaRt** such as getGO.

```
> entrezID = c("983", "3581", "1017")
> rval = getGene(id = entrezID, type = "entrezgene",
```

```
    mart = ensMH)
> unique(rval$hgnc_symbol)

[1] "CDK2" "IL9R" "CDC2"
```

Exercise 8.12
What other data were returned by the call to getGene*?*

In order to obtain other information on the quantities of interest, the getBM function provides a very general interface. In the code below we show how to obtain Interpro domains for the same set of query genes as we used above.

```
> ensembl = useMart("ensembl",
      dataset = "hsapiens_gene_ensembl")
>     ipro = getBM(attributes=c("entrezgene","interpro",
      "interpro_description"),
   filters = "entrezgene", values = entrezID,
      mart = ensembl)
> ipro

   entrezgene  interpro
1        1017 IPR000719
2        1017 IPR008271
3        1017 IPR001245
4        1017 IPR008351
5        1017 IPR002290
6        3581 IPR003531
7         983 IPR000719
8         983 IPR001245
9         983 IPR002290
10        983 IPR008271
                             interpro_description
1                              Protein kinase, core
2     Serine/threonine protein kinase, active site
3                         Tyrosine protein kinase
4                                  JNK MAP kinase
5                 Serine/threonine protein kinase
6         Short hematopoietin receptor, family 1
7                              Protein kinase, core
8                         Tyrosine protein kinase
9                 Serine/threonine protein kinase
10 Serine/threonine protein kinase, active site
```

8.6.4 Getting data from GEO

The Gene Expression Omnibus (GEO) is a repository for gene expression or molecular abundance data. The repository has an online interface where users can select and download data sets of interest. The **GEOquery** package provides a useful set of interface tools that support downloading of GEO data and their conversion into *ExpressionSet* and other Bioconductor data structures suitable for analysis.

The main function in that package is `getGEO`, which is invoked with the name of the data set that you would like to download. It may be advantageous to use the `destdir` argument to store the downloaded file in a permanent location on your local file system as the default location is removed when the R session ends. In the code below, we download a GEO data set and then convert it into an expression set.

```
> library(GEOquery)
> gds = getGEO("GDS1")

File stored at:
/tmp/RtmpvORzTT/GDS1.soft

> eset = GDS2eSet(gds, do.log2 = TRUE)

File stored at:
/tmp/RtmpvORzTT/GPL5.soft
```

The conversion to an *ExpressionSet* is quite complete and all reporter and experiment information is copied into the appropriate locations, as is shown in the example below.

```
> s1 = experimentData(eset)
> abstract(s1)
> s1@pubMedIds


Experiment data
  Experimenter name:
  Laboratory:
  Contact information:
  Title: Testis gene expression profile
  URL:
  PMIDs: 11116097
```

Abstract: A 28 word abstract is available. Use 'abstract' me
thod.
notes:
    :
    able_begin
channel_count:
    1
description:
    Adult testis gene expression profile and gene discovery.
Examines testis, whole male minus gonads, ovary and who
le female minus gonads from adult, 12-24 hours post-eclo
sion, genotype y w[67c1].
feature_count:
    3456
order:
    none
platform:
    GPL5
platform_organism:
    Drosophila melanogaster
platform_technology_type:
    spotted DNA/cDNA
pubmed_id:
    11116097
reference_series:
    GSE462
sample_count:
    8
sample_organism:
    Drosophila melanogaster
sample_type:
    RNA
title:
    Testis gene expression profile
type:
    gene expression array-based
update_date:
    Aug 17 2006
value_type:
    count

8.6.5   KEGG

The Kyoto Encyclopedia of Genes and Genomes (Kanehisa and Goto, 2000) provides a great deal of biological and bioinformatic information. Much of it can be downloaded and processed locally, but they also provide a web service that uses the Simple Object Access Protocol (SOAP). This protocol uses XML to structure requests and responses for web service interactions. The SOAP protocol includes rules for encapsulating requests and responses (e.g., rules for specifying addresses, selecting methods or specifying error handling actions), and for encoding complex data types that form parts of requests and responses (e.g., encoding arrays of floating point numbers).

SOAP services are provided in R through the **SSOAP** package, available from the Omegahat project. And the Bioconductor package **KEGGSOAP** provides an interface to some of the data resources provided by KEGG.

In the example below we obtain the genes in the Riboflavin metabolism pathway in *Saccharomyces cerevisiae*. We then compare the online answer with the answer that can be obtained from data in the **KEGG** package.

```
> library("KEGG")
> library("KEGGSOAP")
> KEGGPATHID2NAME$"00740"

[1] "Riboflavin metabolism"

> SoapAns = get.genes.by.pathway("path:sce00740")
> SoapAns

 [1] "sce:YAR071W" "sce:YBL033C" "sce:YBR092C" "sce:YBR093C"
 [5] "sce:YBR153W" "sce:YBR256C" "sce:YDL024C" "sce:YDL045C"
 [9] "sce:YDR236C" "sce:YDR487C" "sce:YHR215W" "sce:YOL143C"
[13] "sce:YPR073C"
```

Notice that the species abbreviation has been prepended to all gene names. We will use `gsub` to remove the prefix. Then we can use `setdiff` to see if there are any differences between the two.

```
> SA = gsub("^sce:", "", SoapAns)
> localAns = KEGGPATHID2EXTID$sce00740
> setdiff(SA, localAns)

character(0)
```

Chapter 9

Debugging and Profiling

9.1 Introduction

In this chapter we provide some guidance on tools and strategies that should make debugging your code easier and faster. Basically, you must first try to identify the source of the error. While it is generally easy to find where the program actually failed, that is not usually the place where the programming error occurred. Some bugs are reproducible; that is, they occur every time a sequence of commands is executed on all platforms, and others can be more elusive; they arise intermittently and perhaps only under some operating systems. One of the first things that you should do when faced with a likely bug is to try and ensure its reproducibility. If it is not easily reproduced, then your first steps should be to find situations where it is, as only then is there much hope of finding the problem.

One of the best debugging strategies is to write code so that bugs are less likely to arise in the first place. You should prefer the use of simple short functions, each performing a particular task. Such functions are easy to understand and errors are often obvious. Long, convoluted functions tend to both give rise to more bugs and to be more difficult to debug.

This chapter is divided into several sections. First we discuss the `browser` function, which is the main tool used in debugging code in R. R functions such as `debug`, `trace` and `recover` make use of `browser` as a basic tool. The debugging tools are all intended primarily for interactive use and most require some form of user input. We then discuss debugging in R, beginning by recommending that static code analysis using functions from the **codetools** package be used, and then covering some of the basic tools that are available in R. Then we cover debugging procedures that can be applied to detect problems with underlying compiled code. We conclude by discussing tools and methods for profiling memory and the execution of R functions.

9.2   The browser function

The browser function is the building block for many R debugging techniques. A call to browser halts evaluation and starts a special interactive session where the user can inspect the current state of the computations and step through the code one command at a time. The browser can be called from inside any function, and there are ways to invoke the browser when an error or other exception is raised.

Once in the browser, users can execute any R command; they can view the local environment by using ls; and they can set new variables, or change the values assigned to variables simply by using the standard methods for assigning values to variables. The browser also understands a small set of commands specific to it. A summary of the available browser commands and other useful R commands are given in Table 9.1 and Table 9.2. Of these, perhaps the most important to remember for new users is Q, which causes R to quit the debugger and to return control to the command line. Any user input is first parsed to see if it is consistent with a special debugger instruction and, if so, the debugger instruction will be performed. Most of these commands consist of a single letter and are described below. Any local variable with the same name as one of these commands cannot be viewed by simply typing its name, as is standard practice in R, but rather will need to be wrapped in a call to print.

ls()	list the variables defined inside the function
x	print the value of variable x
print(x)	print the value of variable x – useful when x is one of n, l, Q or cont
where	print the call stack
Q	stop the current execution and return to the top-level R interpreter prompt

Table 9.1: Browser commands with non-modal functionalities.

When the browser is active, the prompt changes to Browse[i]> for some positive integer i. The browser can be invoked while a browser session is active, in which case the integer is incremented. Any subsequent calls to browser are nested and control returns to the previous level once a session has

	Initial Mode	Step through Debugger Mode
n	start the step through debugger	execute the next step in the function
c	continue execution	continue execution; if inside a loop, execute until the loop ends
cont	same as c	same as c
carriage return	same as c	same as n

Table 9.2: Browser commands with modal functionalities.

finished. Currently the browser only provides access to the active function; there are no easy ways to investigate the evaluation environments of other functions on the call stack. The browser command `where` can be used to print out the current call stack. To change evaluation environments, you can use direct calls to `recover` from inside of the debugger, but be warned that the set of selections offered may be confusing since for this usage many of the active functions relate to the operation of the browser and not to the evaluation of the function you are interested in.

9.2.1 A sample browser session

Here we show a sample browser session. We first modify the function `setVNames` from the **RBioinf** package so that it starts with a call to browser.

```
> setVNames = function(x, nm) {
+     browser()
+     names(x) = nm
+     asSimpleVector(x, "numeric")
+ }
```

Then, when `setVNames` is invoked, as is shown below, the evaluation of the function call `browser()` halts the execution at that point and a prompt for the browser is printed in the console.

```
> x = 1:10
> x = setVNames(x, letters[1:10])
Browse[1]>
```

At the `browser` prompt, the user can type and execute almost any valid R expression, with the exception of the browser commands described in Tables 9.1 and 9.2, which, if used, will have the interpretation described there.

Sometimes, the user may unintentionally start a large number of nested browser sessions. For example, if the prompt is currently `Browse[2]>`, then the user is at *browser* level 2. Typing c at the prompt will generally continue evaluation of that expression until completion, at which point the user is back at browser level 1 and the prompt will change to `Browse[1]>`. Typing Q will exit from the browser; no further expressions will be evaluated and the user is returned to the top-level R interpreter, where the prompt is >.

9.3 Debugging in R

In this section we describe methods that can be used to debug code that is written in R. As described in the introduction, an important first step is to use tools for static code analysis to try and detect bugs while developing software, rather than at runtime. One aspect of carefully investigating your code for unforeseen problems is the use of the functionality embodied in the **codetools** package. The tools basically inspect R functions and packages and ascertain which variables are local and which are global. They can be used to find variables that are never used or that have no local or global binding, and hence are likely to cause errors.

In the example below, we define a function, `foo`, that we use to demonstrate the use of the **codetools** package by finding all the global variables referenced in the function.

```
> foo = function(x, y) {
+     x = 10
+     z = 20
+     baz(100)
+ }
> library("codetools")
> findGlobals(foo)

[1] "="    "baz" "{"
```

findGlobals reports that there are three global symbols in this function: =, { and baz. The symbols x and y are formal arguments, and hence not global symbols. The numbers, 10, 20 and 100, are constants and hence not symbols, either local or global. And z is a local variable, since it is defined and assigned a value in the body of foo.

In the next code chunk we can find the local variables in the body of the function foo.

```
> findLocals(body(foo))

[1] "x" "z"
```

Notice that x is reported as a local variable, even though it is an argument to foo. The reason is that it is assigned to in the body so that the argument, if supplied, is ignored; and if the argument is not supplied, then x will indeed be local.

The functions that you are likely to use the most are checkUsage and checkUsagePackage. The first checks a single function or closure while the latter checks all functions within the specified package. In the code below, we run checkUsage on the function foo, defined above. Note that the fact that there is no definition for baz is detected as is the fact that z is created but does not seem to be used.

```
> checkUsage(foo, name = "foo", all = TRUE)

foo: no visible global function definition for 'baz'
foo: parameter 'x' changed by assignment
foo: parameter 'y' may not be used
foo: local variable 'z' assigned but may not be used
```

Making use of the tools provided in the **codetools** package can help find a number of problems with your code and using it is well worth the effort. The package checking code, R CMD check, uses **codetools** and reports potential issues.

9.3.1 Runtime debugging

When an error, or unintended outcome, occurs while the program is running, the first step is to locate the source of the error and this is often done in two stages. First you must locate where R has detected the error, and then usually look back from that point to determine where the problem actually occurred. One might think that the important thing is to know which line of which function gave rise to the error. But in many cases, the error arises not because of that particular line, but rather because of some earlier manipulation of the data that rendered it incorrect. Hence, it is often helpful to know which functions are *active* at the time the error was thrown; by active we mean that the body of the function is being evaluated. In R (and most other computer languages), when a function is invoked, the statements in the body of the function are evaluated sequentially. Since each of those statements typically involves one or more calls to other functions, the set of functions that is being evaluated simultaneously can be quite large. When an error occurs, we would like to see a listing of all active functions, generally referred to as the call stack.

While our emphasis, and that of most users, is on dealing with errors that arise, the methods we describe here can be applied to other types of exceptions, such as warnings, which we discuss in Section 9.3.2. But some tools, such as traceback, are specific to errors.

The variable .Traceback stores the call stack for the last uncaught error. Errors that are caught using try or tryCatch do not modify .Traceback. By default, traceback prints the value in .Traceback in a somewhat more user-friendly way. Consider the example below, which makes use of functions supplied in the **RBioinf** package.

```
> x = convertMode(1:4, list())

Error in asSimpleVector(from, mode(to)) : invalid mode list
```

```
> traceback()

3: stop("invalid mode ", mode)
2: asSimpleVector(from, mode(to))
1: convertMode(1:4, list())
```

Each line starting with a number in the output from traceback represents a new function call. Because of lazy evaluation, the sequence of function calls can sometimes be a little odd. Since line numbers are not given, it is not always clear where the error occurred, but at least the user has some sense of which calls were active, and that can greatly help to narrow down the potential causes of the error.

9.3.2   Warnings and other exceptions

Sometimes, instead of getting an error, we get an unexpected warning. Just like unexpected errors, we want to know where they occurred. There are two strategies that you can use. First, you can turn all warnings to errors by setting the `warn` option, as is done in the example below.

```
> saveopt = options(warn = 2)
```

Now any warning will be turned into an error. Later you can restore the settings using the value that was saved when the option was set.

```
> options(saveopt)
```

The second strategy is to use the function `withCallingHandlers`, which provides a very general mechanism for catching errors, warnings, or other conditions and invoking different R functions to debug them. In the example below, we handle warnings; other exceptions can be included by simply adding handlers for them to the call to `withCallingHandlers`.

```
> withCallingHandlers(expression,
+      warning=function(c) recover())
```

9.3.3 Interactive debugging

There are a number of different ways to invoke the browser. Users can have control transferred to the browser on error, they can have the browser invoked on entry to a specific function, and more generally the `trace` function provides a number of capabilities for monitoring when functions are entered or exited. Both `debug` and `trace` interact fairly gracefully with name spaces. They allow the user to debug or trace evaluation within the name space and do not require editing of the source code and rebuilding the package, and so are generally the preferred methods of interacting with code in packages with name spaces.

9.3.3.1 Entering the browser on error

By setting the `error` option, users can request that the browser be invoked when an error is signaled. This can be much simpler than editing the code and placing direct calls to the browser function in the code. In the code chunk below, we can set the `error` option to the function `recover`.

```
> options(error = recover)
```

From this point onwards, until you reset the error option, whenever an error is thrown, R will call the function recover with no arguments. When called, the recover function prints a listing of the active calls and asks the user to select one of them. On selection of a particular call, R starts a browser session inside that call. If the user exits the browser session by typing c, she is again asked to select a call. At any time when making the call selection, the user can return to the R interpreter prompt by selecting 0.

Here is an example session with recover:

```
> x = convertMode(1:4, list())

Error in asSimpleVector(from, mode(to)) : invalid mode list

Enter a frame number, or 0 to exit
1:convertMode(1:4, list())
2:asSimpleVector(from, mode(to))

Selection: 2

Called from: eval(expr, envir, enclos)

Browse[1]> ls()

[1] "mode" "x"

Browse[1]> mode

[1] "list"

Browse[1]> x

[1] 1 2 3 4

Browse[1]> c


Enter a frame number, or 0 to exit
1:convertMode(1:4, list())
2:asSimpleVector(from, mode(to))

Selection: 1

Called from: eval(expr, envir, enclos)
```

```
Browse[1]> ls()

[1] "from" "to"

Browse[1]> to

list()

Browse[1]> Q
```

9.3.4 The debug and undebug functions

It is sometimes useful to enter the browser whenever a particular function is invoked. This can be achieved using the debug function. We will again use the setVNames function, which must first be restored to its original state; this can be done by removing the copy from your workspace, so that the one in the **RBioinf** package will again be found.

```
> rm("setVNames")
```

Then we execute the code below, testing to see if we managed to set the names as intended.

```
> x = matrix(1:4, nrow = 2)
> names(setVNames(x, letters[1:4]))

NULL
```

We see that the names have not been set. Notice also that there is no *error*, but our program is not performing as we would like it to. We suspect the error is in asSimpleVector. So we can apply the function debug to it. This function does nothing more than set a flag on the function that requests that the debugger be entered whenever the function supplied as an argument is invoked.

```
> debug(asSimpleVector)
```

Now any call to asSimpleVector, either directly from the command line or from another function, will start a browser session at the start of the call to asSimpleVector in the step-through debugging mode.

```
> names(setVNames(x, letters[1:4]))

debugging in: asSimpleVector(x, "numeric")
debug: {
    if (!(mode %in% c("logical", "integer", "numeric",
        "double", "complex", "character")))
        stop("invalid mode ", mode)
    Dim = dim(x)
    nDim = length(Dim)
    Names = names(x)
    if (nDim > 0)
        DimNames = dimnames(x)
    x = as.vector(x, mode)
    names(x) = Names
    if (nDim > 0) {
        dim(x) = Dim
        dimnames(x) = DimNames
    }
    x
}

Browse[1]> where

where 1: asSimpleVector(x, "numeric")
where 2: setVNames(x, letters[1:4])

Browse[1]>

debug: if (!(mode %in% c("logical", "integer", "numeric",
    "double", "complex", "character"))) stop("invalid mode ",
    mode)

Browse[1]> x

      [,1] [,2]
[1,]    1    3
[2,]    2    4
attr(,"names")
[1] "a" "b" "c" "d"
```

As we suspected, at entry, the parameter x has the names attribute set. So the error must be somewhere inside this function. We continue the debugging and examining the value of x.

```
Browse[1]>

debug: Dim = dim(x)

Browse[1]>

debug: nDim = length(Dim)

Browse[1]> Dim

[1] 2 2

Browse[1]>

debug: Names = names(x)

Browse[1]> nDim

[1] 2

Browse[1]>

debug: if (nDim > 0) DimNames = dimnames(x)

Browse[1]> Names

[1] "a" "b" "c" "d"

Browse[1]>

debug: x = as.vector(x, mode)

Browse[1]>

debug: names(x) = Names

Browse[1]> x

[1] 1 2 3 4

Browse[1]>

debug: if (nDim > 0) {
    dim(x) = Dim
    dimnames(x) = DimNames
}

Browse[1]> x

a b c d
1 2 3 4
```

We have correctly set the value of x back.

```
Browse[1]>

debug: dim(x) = Dim

Browse[1]>

debug: dimnames(x) = DimNames

Browse[1]> x

      [,1] [,2]
[1,]    1    3
[2,]    2    4
```

However, after setting the dimension, the names attribute gets removed. Now we know where the error is — we should set the name attribute after setting the dimension and the dimnames. We first go to the end of the function.

```
Browse[1]>

debug: dimnames(x) = DimNames

Browse[1]> x

      [,1] [,2]
[1,]    1    3
[2,]    2    4

Browse[1]>

debug: x
```

Then we verify that setting the names does not disturb the dimension and then quit from the browser.

```
Browse[1]> names(x) = Names
Browse[1]> x

      [,1] [,2]
[1,]    1    3
```

```
[2,]    2    4
attr(,"names")
[1] "a" "b" "c" "d"

Browse[1]> Q
```

After finishing debugging, we undebug asSimpleVector, and now the debugger will not be called on entry to asSimpleVector.

```
> undebug(asSimpleVector)
```

There is no easy way to find out which functions are currently being debugged.

9.3.5   The trace function

The trace function provides all the functionality of the debug function and it can do some other useful things. First of all, it can be used to just print all calls to a particular function when it is entered and exited.

```
> trace(asSimpleVector)
> x = list(1:3, 4:5)
> for (i in seq(along = x)) {
+     x[[i]] = asSimpleVector(x[[i]], "complex")
+ }

trace: asSimpleVector(x[[i]], "complex")
trace: asSimpleVector(x[[i]], "complex")

> untrace(asSimpleVector)
```

Each time the function being traced is called, a line is printed starting with trace: and followed by the call. Here the asSimpleVector function was called twice inside the for loop. That is why we see two lines starting with trace:. A call to untrace stops the tracing.

Secondly, it can be used like debug — but to only start the browsing at a particular point inside the function. Suppose we want to start the browser just before we enter the if block that sets the dimension and the dimnames. We can use the function printWithNumbers to print asSimpleVector with appropriate line numbers, the index of that place in the function. The function is printed in the code chunk below and break points can be set for any line that has a number. When set, the `tracer` function will be evaluated just prior to the evaluation of the specified line number.

```
> printWithNumbers(asSimpleVector)


    function (x, mode = "logical")
1: {
2:      if (!(mode %in% c("logical", "integer", "numeric", "dou
ble",
            "complex", "character")))
            stop("invalid mode ", mode)
3:      Dim <- dim(x)
4:      nDim <- length(Dim)
5:      Names <- names(x)
6:      if (nDim > 0)
            DimNames <- dimnames(x)
7:      x <- as.vector(x, mode)
8:      names(x) <- Names
9:      if (nDim > 0) {
            dim(x) <- Dim
            dimnames(x) <- DimNames
        }
10:     x
    }
    <environment: namespace:RBioinf>
```

By default, a call to trace prints the call. We can make it call `browser` by supplying the `tracer` argument. We can start the tracing at a particular place inside the function by supplying the `at` argument. To start tracing at the beginning of the `if` block setting the dimension, we used `at=9` in our call to `trace`.

```
> trace(asSimpleVector, tracer = browser, at = 9)

[1] "asSimpleVector"
```

And now when the debugger is invoked at the line number requested, all statements above that one have been evaluated and users can query and modify values, as for any other invocation of the browser.

```
> names(setVNames(1:4, letters[1:4]))

Tracing asSimpleVector(x, "numeric") step 9
Called from: asSimpleVector(x, "numeric")

Browse[1]> ls()

[1] "Dim"    "Names" "mode"   "nDim"   "x"

Browse[1]> x

a b c d
1 2 3 4

Browse[1]> Q
```

We halt tracing by calling untrace with the function we want to stop tracing as an argument.

```
> untrace(asSimpleVector)
```

Finally, the trace function can also be used to debug calls to a particular method for a S4 generic function (Section 3.7). To demonstrate that, we turn the subsetAsCharacter function into an S4 generic function.

```
> setGeneric("subsetAsCharacter")

[1] "subsetAsCharacter"
```

In addition to creating a generic from the existing subsetAsCharacter function, this command also sets the original function as the default method. We define an additional method for character vectors and simple subscripts.

```
> setMethod("subsetAsCharacter", signature(x = "character",
      i = "missing", j = "missing"), function(x,
      i, j) x)

[1] "subsetAsCharacter"
```

Now we will use `trace` to debug the `subsetAsCharacter` generic only when x is of class `"character"`.

```
> trace("subsetAsCharacter", tracer = browser,
      signature=c(x = "numeric"))

[1] "subsetAsCharacter"
```

Note that, in this particular case, there was no specific `subsetAsCharacter` method with this signature. So the tracing will occur for the default method — but only when the signature matches the one given to `trace`.

```
> subsetAsCharacter(1.5, 1:2)

Tracing subsetAsCharacter(1.5, 1:2) on entry
Called from: subsetAsCharacter(1.5, 1:2)

Browse[1]> ls()

[1] "i" "j" "x"

Browse[1]> x

[1] 1.5

Browse[1]> c

[1] "1.5" NA
```

```
> subsetAsCharacter(1 + (0+0i), 1:2)

[1] "1+0i" NA

> subsetAsCharacter("x")

[1] "x"

> untrace("subsetAsCharacter")
```

9.4 Debugging C and other foreign code

Debugging compiled code is quite complex and generally requires some knowledge of programming, how compiled programs are evaluated and other rather esoteric details. In this section we presume a fairly high standard of knowledge and recommend that if you have not used any of the tools described here, or similar tools, you should consider consulting a local expert for advice and guidance. URLs are given for the different software discussed, and readers are referred to those locations for complete documentation of the tools presented. The R Extensions Manual also provides some more detailed examples and discussions that readers may want to consult.

The most widely used debugger for compiled code is gdb (see http://www.gnu.org/software/gdb). It can be used on Windows (provided you have installed the tools for building and compiling your own version of R and R packages), Unix, Linux and OS X. The ddd(http://www.gnu.org/software/ddd/) graphical interface to gdb can be quite helpful for users not familiar with gdb.

In order to make use of gdb, you must compile R, and all compiled code that you want to inspect, using the appropriate compiler flags. The compiler flags can be set in the file R_HOME/config.site. We suggest turning off all optimization; to yield the best results, do not use -O2 or similar, and use the -g flag . While gdb is supposed to be able to deal with optimized compiled code, there are often small glitches, and using no optimization removes this potential source of confusion.

If you change these flags, you will need to remake all of R, typically by issuing the make clean directive, followed by make. Any libraries that have been installed and that have source code will need to have the source recompiled using the new flags, if you intend to debug them.

R can be invoked with the ddd debugger by using the syntax R -d ddd, or equivalently R --debugger=ddd. Similar syntax is used for other debuggers. Options can be passed through to the debugger by using the --debugger-args option as well.

Unix-like systems can make use of valgrind (http://valgrind.org) to check for memory leaks and other memory problems. The code given below runs valgrind while evaluating the code in the file someCode.R. Valgrind can make your code run quite slowly, so be patient when using it.

```
R -d "valgrind --tool=memcheck --leak-check=yes"
    --vanilla < someCode.R
```

9.5   Profiling R code

There are often situations where code written in R takes rather a long time to run. In very many cases, the problem can be overcome simply by making use of more appropriate tools in R, by rearranging the code so that the computations are more efficient, or by vectorizing calculations. In some cases, when even after all efforts have been expended, the code is still too slow to be viable, rewriting parts of the code in C or some other foreign language (see Chapter 6 for more complete details) may be appropriate. However, in all cases, it is still essential that a correct diagnosis of the problem be made. That is, it is essential to determine which computations are slow and in need of improvement. This is especially important when considering writing code in a compiled language, since the diagnosis can help to greatly reduce the amount of foreign code that is needed and in some cases can help to identify a particular programming construct that might valuably be added to R itself.

Another tool that is often used is timing comparison. That is, two different implementations are run and the time taken for each is recorded and reported. While this can be valuable, some caution in interpreting results is needed. Since R carries out its own memory management, it is possible that one version will incur all of the costs of memory allocation and hence look much slower.

The functions `Rprof` and `summaryRprof` can be used to profile R commands and to provide some insight into where the time is being spent. In the next code chunk, we make use of `Rprof` to profile the computation of the median absolute deviation about the median (or MAD) on a large set of simulated data. The first call to `Rprof` initiates profiling. `Rprof` takes three optional arguments: first the name of the file to print the results to, second a logical argument indicating whether to overwrite or append to the existing file, and third the sampling interval, in seconds. Setting this too small, below what the operating system supports, will lead to peculiar outputs. We make use of the default settings in our example.

```
> Rprof()
> mad(runif(1e+07))

[1] 0.371

> Rprof(NULL)
```

The second call to `Rprof`, with the argument `NULL`, turns profiling off. The contents of the file `Rprof.out` are the active calls, computed every `interval` seconds. These can be summarized by a call to `summaryRprof`, which tabulates them and reports on the time spent in different functions.

291

```
> summaryRprof()

$by.self
                self.time self.pct total.time total.pct
"sort.int"          0.20     35.7       0.24      42.9
"is.na"             0.14     25.0       0.14      25.0
"runif"             0.10     17.9       0.10      17.9
"-"                 0.06     10.7       0.06      10.7
"abs"               0.04      7.1       0.04       7.1
"list"              0.02      3.6       0.02       3.6
"<Anonymous>"       0.00      0.0       0.56     100.0
"Sweave"            0.00      0.0       0.56     100.0
"doTryCatch"        0.00      0.0       0.56     100.0
"evalFunc"          0.00      0.0       0.56     100.0
"try"               0.00      0.0       0.56     100.0
"tryCatch"          0.00      0.0       0.56     100.0
"tryCatchList"      0.00      0.0       0.56     100.0
"tryCatchOne"       0.00      0.0       0.56     100.0
"eval.with.vis"     0.00      0.0       0.54      96.4
"mad"               0.00      0.0       0.54      96.4
"median"            0.00      0.0       0.44      78.6
"median.default"    0.00      0.0       0.34      60.7
"mean"              0.00      0.0       0.24      42.9
"sort"              0.00      0.0       0.24      42.9
"sort.default"      0.00      0.0       0.24      42.9

$by.total
                total.time total.pct self.time self.pct
"<Anonymous>"         0.56     100.0      0.00      0.0
"Sweave"              0.56     100.0      0.00      0.0
"doTryCatch"          0.56     100.0      0.00      0.0
"evalFunc"            0.56     100.0      0.00      0.0
"try"                 0.56     100.0      0.00      0.0
"tryCatch"            0.56     100.0      0.00      0.0
"tryCatchList"        0.56     100.0      0.00      0.0
"tryCatchOne"         0.56     100.0      0.00      0.0
"eval.with.vis"       0.54      96.4      0.00      0.0
"mad"                 0.54      96.4      0.00      0.0
"median"              0.44      78.6      0.00      0.0
"median.default"      0.34      60.7      0.00      0.0
"sort.int"            0.24      42.9      0.20     35.7
"mean"                0.24      42.9      0.00      0.0
"sort"                0.24      42.9      0.00      0.0
"sort.default"        0.24      42.9      0.00      0.0
```

```
"is.na"                 0.14        25.0        0.14        25.0
"runif"                 0.10        17.9        0.10        17.9
"-"                     0.06        10.7        0.06        10.7
"abs"                   0.04         7.1        0.04         7.1
"list"                  0.02         3.6        0.02         3.6

$sampling.time
[1] 0.56
```

The output has three components. There are two arrays, the first sorted by self-time and the second sorted by total-time. The third component of the response is the total time spent in the execution of the commands.

Given the command, it is no surprise that all of the total-time was spent in the function `mad`. However, since the self-time for that function is zero, we can conclude that computational effort was expended elsewhere. When looking at self-time, we see that the bulk of the time is spent in `sort.int`, `runif` and `is.na`. And, since we know that there are no missing values, it does seem that some savings are available, as there is no need to run the `is.na` function. Although one is able to control checking for `NAs` in the call to `mad`, no such fine-grained control is possible with `sort`. Hence, you must either live with the inefficiency or write your own version of `sort` that does allow the user to turn off checking for missing values.

9.5.1 Timings

The basic tool for timing is `system.time`. This function returns a vector of length five, but only three of the values are normally printed. The three elements are the user cpu time, system cpu time, and elapsed time. Times are reported in seconds, the resolution is system specific, but is typically to 1/100th of a second.

In the output shown below, the same R code was run three times, simultaneously, in a pristine R session. As you can see, there is about a 5% difference between the system time for the first evaluation and those of the subsequent evaluations. So when comparing the execution time of different methods, it is prudent to change the order, and to repeat the calculations in different ways, to ensure that the observed effects are real and important.

```
>   system.time(mad(runif(10000000)))

   user   system  elapsed
  1.821    0.663    2.488

>   system.time(mad(runif(10000000)))
```

```
   user   system elapsed
   1.817  0.635   2.455

> system.time(mad(runif(10000000)))

   user   system elapsed
   2.003  0.632   2.638
```

The optional argument `gcFirst` is `TRUE` by default and ensures that R's garbage collector is run prior to the evaluation of the supplied expression. By running the garbage collector first, it is likely that more consistent timings will be produced.

9.6 Managing memory

There are some tools available in R to monitor memory usage. In R, memory is divided into two separate components: memory for atomic vectors (e.g., integers, characters) and language elements. The language elements are the SEXPs described in Chapter 6, while vector storage is contiguous storage for homogeneous elements. Vector storage is further divided into two types: the small vectors, currently less than 128 bytes, which are allocated by R (which obtains a large chunk of memory, and then parcels it out as needed) and larger vectors for which memory is obtained directly from the operating system.

R attempts to manage memory effectively and has a generational garbage collector. Explicit details on the functioning of the garbage collector are given in the R Internals manual (R Development Core Team, 2007d). During normal use, the garbage collector runs automatically whenever storage requests exceed the current free memory available. A user can trigger garbage collection with the `gc` command, which will report the number of `Ncells` (SEXPs) used and the number of `Vcells` (vector storage) used, as well as a few other statistics. The function `gcinfo` can be used to have information print every time the garbage collector runs.

```
> gc()

          used (Mb) gc trigger (Mb) max used  (Mb)
Ncells  318611  8.6     597831   16   407500  10.9
Vcells  165564  1.3   29734436  227 35230586 268.8
```

One can also find out how many of the `Ncells` are allocated to each of the different types of SEXPs using `memory.profile`. In the example below, we obtain the output of `memory.profile` and sort it, from largest to smallest. This should be approximately equal to the value for `Ncells` used by `gc`, but minor discrepancies are likely to occur to reflect the creation of new objects or the effects of garbage collection.

```
> ss = memory.profile()
> sort(ss, decreasing = TRUE)

   pairlist    language   character        char      symbol
     176473       48112       41736        9838        7242
    integer        list     logical     promise     closure
       5974        5336        5124        4934        4579
     double     builtin environment          S4 externalptr
       3045        2035        1675        1654         513
    special     weakref     complex  expression        NULL
        224         121           3           2           1
        ...         raw         any    bytecode
          1           1           0           0

> sum(ss)

[1] 318623
```

9.6.1 Memory profiling

Memory profiling has an adverse effect on performance, even if it is not being used, and hence is implemented as a compile time option. To use memory profiling, R must be compiled with it enabled. This means that readers will have to ensure that their version of R has been compiled to allow for memory profiling if they want to follow the examples in this section.

There are three different strategies that can be used to profile memory usage. You can incorporate memory usage information in the output created by `Rprof` by setting the argument `memory.profiling` to `TRUE`. And in that case, information about total memory usage is reported for each sampling time. The information can then be summarized in different ways using `summaryRprof`. There are four options for summarizing the output; `none` (the default) excludes memory usage information, while `both` requests that memory usage information be printed with the other profiling information.

Two more advanced options are `tseries` and `stats`, which require that a second argument, `index`, also be specified. The `index` argument specifies how to summarize the calls on the stack trace. In the code below, we examine memory usage from performing RMA on the Dilution data. First we load the necessary packages, then set up profiling and run the code we want to profile.

```
> library("affy")
> library("affydata")
> data(Dilution)
> Rprof(file = "profRMA", memory.profiling = TRUE)
> r1 = rma(Dilution)

Background correcting
Normalizing
Calculating Expression

> Rprof(NULL)
```

And in the next code segment, we read in the profiling data and display selected parts of it. By setting `memory` to `"tseries"`, the return value is a data frame with one row for each sampling time, and values that help track usage of vector storage (both large and small), language elements (nodes), calls to duplicate, and the call stack at the time the data were sampled.

```
> pS = summaryRprof(file = "profRMA", memory = "tseries")
> names(pS)

[1] "vsize.small"  "vsize.large"  "nodes"
[4] "duplications" "stack:2"
```

Users can then examine these statistics to help identify potential inefficiencies in their code. For example, we plot the number of calls to `duplicate`. What is quite remarkable in this plot is that there are a few spikes in calls to `duplicate`, which are in the thousands. While such duplication may be necessary, it is likely that it is not. Further tracking down the source of this and making sure it is necessary could greatly speed up the processing time and possibly decrease memory usage.

9.6.2 Profiling memory allocation

Another mechanism for memory profiling is provided by the `Rprofmem` function, which collects and prints information on the call stack whenever a large (as determined by the user) object is allocated. The argument `threshold` sets the size threshold, in bytes, for recording the memory allocation. This tool can help to identify inefficiencies that arise due to copying large objects without getting overwhelmed by the total number of copies. As observed in

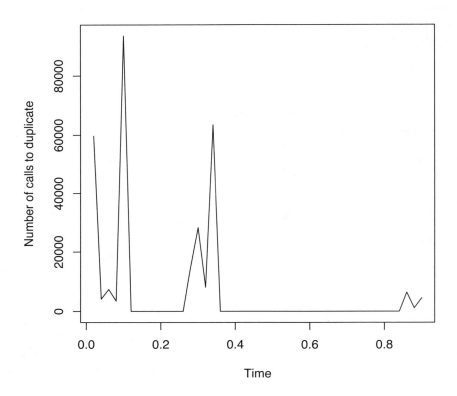

FIGURE 9.1: Time series view of calls to duplicate during the processing of Affymetrix data.

Figure 9.1, there are very many calls to `duplicate` during the evaluation of the `rma` function. It is not clear whether these are large or small objects.

In the next code segment, we request that allocation of objects larger than 10000 bytes be recorded. Once the computations are completed, we view the first five lines of the output file. The functions being called suggest that there is a lot of allocation begin performed to retrieve the probe names. In the example, we needed to trim the output, using `strtrim`, so that it fits on the page; readers would not normally do that.

```
> Rprofmem(file = "rma2.out", threshold = 1e+05)
> s2 = rma(Dilution)

Background correcting
Normalizing
Calculating Expression

> Rprofmem(NULL)


> noquote(readLines("rma2.out", n = 5))


[1] new page:".deparseOpts" "deparse" "eval" "match.arg" ".loc
al" "indexProbes" "indexProbes" ".local" "pmindex" "pmindex" "
.local" "probeNames" "probeNames" "rma" "eval.with.vis" "doTry
Catch" "tryCatchOne" "tryCatchList" "tryCatch" "try" "evalFunc
" "<Anonymous>" "Sweave"

[2] new page:"switch" "<Anonymous>" "data" "cleancdfname" "cdf
FromLibPath" "switch" "getCdfInfo" ".local" "indexProbes" "ind
exProbes" ".local" "pmindex" "pmindex" ".local" "probeNames" "
probeNames" "rma" "eval.with.vis" "doTryCatch" "tryCatchOne" "
tryCatchList" "tryCatch" "try" "evalFunc" "<Anonymous>" "Sweav
e"
[3] new page:"match" "cleancdfname" "cdfFromLibPath" "switch"
"getCdfInfo" ".local" "indexProbes" "indexProbes" ".local" "pm
index" "pmindex" ".local" "probeNames" "probeNames" "rma" "eva
l.with.vis" "doTryCatch" "tryCatchOne" "tryCatchList" "tryCatc
h" "try" "evalFunc" "<Anonymous>" "Sweave"

[4] new page:"file.info" ".find.package" "cdfFromLibPath" "swi
tch" "getCdfInfo" ".local" "indexProbes" "indexProbes" ".local
" "pmindex" "pmindex" ".local" "probeNames" "probeNames" "rma"
 "eval.with.vis" "doTryCatch" "tryCatchOne" "tryCatchList" "tr
```

```
yCatch" "try" "evalFunc" "<Anonymous>" "Sweave"

[5] new page:"as.vector" ".local" "indexProbes" "indexProbes"
".local" "pmindex" "pmindex" ".local" "probeNames" "probeNames
" "rma" "eval.with.vis" "doTryCatch" "tryCatchOne" "tryCatchLi
st" "tryCatch" "try" "evalFunc" "<Anonymous>" "Sweave"
```

```
> length(readLines("rma2.out"))
```

```
[1] 6239
```

Exercise 9.1
Write a function to parse the output of Rmemprof *and determine the total
amount of memory allocated. Use the names from the call stack to assign the
memory allocation to particular functions.*

9.6.3   Tracking a single object

The third mechanism that is provided is to trace a single object and determine when and where it is duplicated. The function tracemem is called with
the object to be traced, and subsequently, whenever the object (or a natural
descendant) is duplicated, a message is printed.

In the code below, we first trace duplication of Dilution in the call to
rma, but find that there is none; and there should be none, so that is good.
When subsetting an instance of the *ExpressionSet* class, however, it seems
that around four copies are made and none should be, so there are definitely
some inefficiencies that could be fixed.

```
> tracemem(Dilution)
```

```
[1] "<0x2e2da40>"
```

```
> s3 <- rma(Dilution)
```

```
Background correcting
Normalizing
Calculating Expression
```

```
> tracemem(s3)
```

```
[1] "<0x8420bc0>"
```

```
>  s2 = s3[1:100,]

tracemem[0x6367894 -> 0x5759994]: [ [
tracemem[0x5759994 -> 0x6517df0]: featureData<-
tracemem[0x6517df0 -> 0x55de304]: [ [
tracemem[0x55de304 -> 0x560156c]: assayData<-
```

Exercise 9.2
Trace memory usage on an instance of the ExpressionSet class when setting the sample names. How many copies are made?

References

H. Abelson and G. J. Sussman. *Structure and Interpretation of Computer Programs*. MIT Press, Cambridge, MA, 2nd edition, 1996.

R. A. Becker, J. M. Chambers, and A. R. Wilks. *The New S Language: A Programming Environment for Data Analysis and Statistics*. Wadsworth, Pacific Grove, CA, 1988.

J. Bentley. *Programming Pearls*. Addison-Wesley, 2nd edition, 1999.

E. Camon, M. Magrane, D. Barrell, et al. The Gene Ontology Annotation (GOA) database: sharing knowledge in Uniprot with Gene Ontology. *Nucleic Acids Research*, 32:D262–D266, 2004.

J. M. Chambers. *Programming with Data: A Guide to the S Language*. Springer-Verlag, New York, 1998.

J. M. Chambers. *Software for Data Analysis: Programming with R*. Springer, New York, 2008.

J. M. Chambers and T. Hastie. *Statistical Models in S*. Wadsworth, Pacific Grove, CA, 1992.

T. Chiang, N. Li, S. Orchard, et al. Rintact: a direct link between molecular interaction data and methods in proteomic analysis. *Bioinformatics*, page doi: 10.1093/bioinformatics/btm518, 2007.

T. Cormen, C. Leiserson, and R. Rivest. *Introduction to Algorithms*. McGraw-Hill, New York, 1990.

S. R. Eddy. Where did the BLOSUM62 alignment score matrix come from? *Nature Biotechnology*, 22(8):1035–1036, 2004.

M. Eisler. XDR: External data representation standard. RFC 4506 (Standard), 2006. URL http://tools.ietf.org/html/rfc4506.

S. Fields and O. Song. A novel genetic system to detect protein-protein interactions. *Nature*, 340:245–246, 1989.

D. P. Freidman, M. Wand, and C. T. Haynes. *Essentials of Programming Languages*. MIT Press, Cambridge, MA, 2nd edition, 2001.

J. E. F. Friedl. *Mastering Regular Expressions*. O'Reilly, Sebastopol, CA, 2nd edition, 2002.

E. Gamma, R. Helm, R. Johnson, and J. Vlissides. *Design Patterns.* Addison-Wesley, Boston, 1995.

R. Gentleman. Reproducible research: a bioinformatics case study. *Statistical Applications in Genetics and Molecular Biology*, 4, 2005. URL http://www.bepress.com/sagmb/vol4/iss1/art2.

R. Gentleman and J. Gentry. Querying PubMed. *R News*, 2(2):28–31, 2002. URL http://CRAN.R-project.org/doc/Rnews.

R. Gentleman and R. Ihaka. Lexical scope and statistical computing. *Journal of Computational and Graphical Statistics*, 9:491–508, 2000.

R. Gentleman and D. Temple Lang. Statistical analyses and reproducible research. *Journal of Computational and Graphical Statistics*, 16:1–23, 2007.

R. C. Gentleman, V. J. Carey, D. M. Bates, et al. Bioconductor: open software development for computational biology and bioinformatics. *Genome Biology*, 5:R80, 2004. URL http://genomebiology.com/2004/5/10/R80.

D. Goldberg. What every computer scientist should know about floating-point arithmetic. *ACM Computing Surveys*, 23:5–48, 1991.

D. Gusfield. *Algorithms on Strings, Trees and Sequences.* Cambridge University Press, New York, 1997.

F. Hahne, W. Huber, R. Gentleman, and S. Falcon. *Bioconductor Case Studies.* Springer, New York, 2008.

E. R. Harold and W. S. Means. *XML in a Nutshell.* O'Reilly, Sebastopol, CA, 3rd edition, 2004.

B. Haubold and T. Wiehe. *Introduction to Computational Biology, An Evolutionary Approach.* Birkhäuser, Basel, 2006.

S. Kamin. *Programming Languages: An Interpreter-Based Approach.* Addison-Wesley, Boston, 1990.

M. Kanehisa and S. Goto. KEGG: Kyoto encyclopedia of genes and genomes. *Nucleic Acids Research*, 28:27–30, 2000.

B. W. Kernighan and D. M. Ritchie. *The C Programming Language.* Prentice Hall, New York, 2nd edition, 1988.

S. Kerrien et al. Intact – open source resource for molecular interaction data. *Nucleic Acids Research*, 35:D561–D565, 2006.

S. Kurtz, A. Phillippy, A. L. Delcher, et al. Versatile and open software for comparing large genomes. *Genome Biology*, page 5:R12, 2004. URL http://genomebiology.com/2004/5/2/R12.

K. Lange. *Numerical Analysis for Statisticians.* Springer, New York, 1999.

F. Leisch. Sweave: dynamic generation of statistical reports using literate data analysis. In W. Härdle and B. Rönz, editors, *Compstat 2002 — Proceedings in Computational Statistics*, pages 575–580. Physika Verlag, Heidelberg, Germany, 2002. URL http://www.ci.tuwien.ac.at/ leisch/Sweave. ISBN 3-7908-1517-9.

P. Murrell. *R Graphics.* Chapman & Hall/CRC, New York, 2005.

R Development Core Team. *R Data Import/Export.* R Foundation for Statistical Computing, Vienna, Austria, 2007a. URL http://www.R-project.org. ISBN 3-900051-10-0.

R Development Core Team. *R Language Definition.* R Foundation for Statistical Computing, Vienna, Austria, 2007b. URL http://www.R-project.org. ISBN 3-900051-13-5.

R Development Core Team. *Writing R Extensions.* R Foundation for Statistical Computing, Vienna, Austria, 2007c. URL http://www.R-project.org. ISBN 3-900051-11-9.

R Development Core Team. *R Internals.* R Foundation for Statistical Computing, Vienna, Austria, 2007d. URL http://www.R-project.org. ISBN 3-900051-14-3.

B. D. Ripley. Lazy loading and packages in R 2.0.0. *R News*, 4:2–4, 2004.

D. Sarkar. *Lattice: Multivariate Data Visualization with R.* Springer, New York, 2008.

R. Sedgewick. *Algorithms in C.* Addison-Wesley, Boston, 2001.

A. Shalit. *The Dylan Reference Manual.* Apple Press, 1996.

A. Skonnard and M. Gudgin. *Essential XML Quick Reference.* Addison-Wesley, Boston, 2001.

G. L. Steele. *Common LISP The Language.* Digital Press, Woburn, MA, 2nd edition, 1990.

W. R. Stevens and S. A. Rago. *Advanced programming in the UNIX environment.* Addison-Wesley, Boston, 2nd edition, 2005.

T. Stubblebine. *Regular Expression Pocket Reference.* O'Reilly, Sebastopol, CA, 2nd edition, 2007.

The Gene Ontology Consortium. Gene Ontology: tool for the unification of biology. *Nature Genetics*, 25:25–29, 2000.

R. Thisted. *Elements of Statistical Computing.* Chapman & Hall/CRC, New York, 1988.

L. Tierney. Name space management for R. *R News*, 3(1):2–5, 2003. URL http://CRAN.R-project.org/doc/Rnews.

L. Tierney. Simple references with finalization. Technical report, 2002. URL http://www.stat.uiowa.edu/ luke/R/simpleref.html.

W. N. Venables and B. D. Ripley. *Modern Applied Statistics with S (4e).* Springer, New York, 2002.

W. N. Venables and B. D. Ripley. *S Programming.* Springer, New York, 2000.

Index

306

308